T0390367

Engendering Development

Engendering Development demonstrates how gender is a form of inequality that is used to generate global capitalist development. It charts the histories of gender, race, class, sexuality and nationality as categories of inequality under imperialism, which continue to support the accumulation of capital in the global economy today.

The textbook draws on feminist and critical development scholarship to provide insightful ways of understanding and critiquing capitalist economic trajectories by focusing on the way development is enacted and protested by men and women. It incorporates analyses of the lived experiences in the global north and south in place-specific ways. Taking a broad perspective on development, *Engendering Development* draws on textured case studies from the authors' research and the work of geographers and feminist scholars. The cases demonstrate how gendered, raced and classed subjects have been enrolled in global capitalism, and how individuals and communities resist, embrace and rework development efforts. This textbook starts from an understanding of development as global capitalism that perpetuates and benefits from gendered, raced and classed hierarchies.

The book will prove to be useful to advanced undergraduate and graduate students enrolled in courses on development through its critical approach to development conveyed with straightforward arguments, detailed case studies, accessible writing and a problem-solving approach based on lived experiences.

Amy Trauger is an Associate Professor in the Department of Geography at the University of Georgia, USA. She has published more than 20 journal articles on gender, labor and sustainability in agriculture, organic food supply chains and food sovereignty. She is the author of *We Want Land to Live! Making Political Space for Life*. She edited *Food Sovereignty in International Context*, published by Earthscan/Routledge, and is a co-editor of *Making Policy for Food Sovereignty: Social Movements, Markets and the State* with Priscilla Claeys and Annette Desmarais, also published by Earthscan/Routledge.

Jennifer L. Fluri is an Associate Professor in the Department of Geography at the University of Colorado-Boulder, USA. She has over 20 publications in peer-reviewed academic journals. Internationally, her research focuses on gender, geopolitics and international development in Afghanistan. In Colorado, she co-directs the Boulder Affordable Housing Research Initiative, a collaborative and service-based research project (Colorado.edu/BAHRI). She has co-authored two books, *Carpetbaggers of Kabul and other American-Afghan Entanglements* with Rachel Lehr, and *Feminist Spaces: Gender and Geography in a Global Context* with Ann Oberhauser, Risa Whiston and Sharlene Mollett. She is the co-editor, with Katharyne Mitchell and Reece Jones, of the *Handbook on Critical Geographies of Migration.*

Engendering Development

Capitalism and Inequality in the Global Economy

Amy Trauger and Jennifer L. Fluri

LONDON AND NEW YORK

First published 2019
by Routledge
2 Park Square, Milton Park, Abingdon, Oxon OX14 4RN

and by Routledge
52 Vanderbilt Avenue, New York, NY 10017

Routledge is an imprint of the Taylor & Francis Group, an informa business

British Library Cataloguing-in-Publication Data
A catalogue record for this book is available from the British Library

Library of Congress Cataloging-in-Publication Data
Names: Trauger, Amy, author. | Fluri, Jennifer L., author.
Title: Engendering development : capitalism and inequality in the global economy / Amy Trauger and Jennifer L. Fluri.
Description: Abingdon, Oxon ; New York, NY : Routledge, 2019. | Includes bibliographical references and index.
Subjects: LCSH: Women in development. | Economic development. | Capitalism. | Equality.
Classification: LCC HQ1240 .T74 2019 | DDC 305.42—dc23
LC record available at https://lccn.loc.gov/2018060398

ISBN: 978-0-415-78966-0 (hbk)
ISBN: 978-0-415-78967-7 (pbk)
ISBN: 978-1-315-21384-2 (ebk)

Typeset in Minion Pro
by Apex CoVantage, LLC

We live in capitalism. Its power seems inescapable. So did the divine right of kings. Any human power can be resisted and changed by human beings. Resistance and change often begin in art, and very often in our art, the art of words.

Ursula K. Le Guin

Contents

List of illustrations ix
Acknowledgements xi

PART I UNDERSTANDING DEVELOPMENT AND INEQUALITY 1

1 Understanding development and inequality 3

2 Engendering development 19

3 The business of international development 35

PART II PROCESSES IN DEVELOPMENT 49

4 Development as dispossession 51

5 Labor, migration and capital accumulation 66

6 Work, mobility and uneven development 81

PART III MOMENTS IN DEVELOPMENT 93

7 Health and population 95

8 Gender and development technologies 112

9 Disaster assistance and development 127

10 Alternative development and decolonization 141

Index 153

Illustrations

Figures

2.1 Map of Africa pre-contact 20
2.2 Jantar Mantar 24
3.1 Fair trade bananas in the Dominican Republic 45
4.1 Proposed route of DAPL 52
4.2 Rural land use in the Dominican Republic 57
5.1 Rural Garhwali migrant returning home for a visit 67
5.2 Women's hospital in Havana, Cuba 71
5.3 Informal farmers' market in Uttarakhand, India 75
5.4 United Arab Emirates population pyramid 77
6.1 Maternity leave map 83
7.1 Rana Plaza building collapse, April 24, 2013, Dhaka, Bangladesh 98
7.2 Japan's population pyramid 101
7.3 Sierra Leone's population pyramid 101
7.4 2011 map of Indian states sex ratio 108
7.5 A small family is a happy family – postage stamp 109
8.1 Maquiladora factory 116
8.2 Biotechnology laboratory in Nairobi, Kenya 122
8.3 Woman working over a coal stove 123
8.4 Biogas diagram 124

Table

6.1 Top ten countries for gender equality, the *Global Gender Gap Report,* 2016 83

Acknowledgements

We would like to thank all the feminist scholars who inspire us, our students who take us to new places intellectually and the people we've worked with all over the world who told us their stories. We are grateful to editor Andrew Mould for his faith in this project and editorial assistant Egle Zigaite for endless patience. We also appreciate very much the anonymous reviewers who provided constructive criticism and feedback on the proposal and earlier drafts of the manuscript.

Cover photo courtesy of Avery Leigh White.

Part I

Understanding development and inequality

1 Understanding development and inequality

Introduction

At the United Nations (UN) General Assembly in 2017, world leaders and other officials gathered to discuss some of the gravest and most difficult challenges facing the world, specifically those facing the world's poorest people. The UN group known as "Project Everyone" seized an opportunity to have a captive audience, and proposed to the Mars Corporation that they develop and donate candy printed with symbols representing seven of the seventeen newly adopted Sustainable Development Goals. These goals include efforts to eliminate poverty and inequality and improve the environmental sustainability of development projects. Chocolate multinational corporations have long been implicated in perpetuating both poverty and inequality, particularly with respect to cocoa extraction in West Africa. This includes the trafficking of child slaves, especially girls as plantation laborers (Dottridge 2002). Reporting on the UN event neglected to mention the source of the cocoa or its effects on workers, focusing rather on bringing a little "fun" to heavy topics (Gharib 2017). Media coverage of this event did not address the way multinational corporations (such as Mars) and their relationship to supranational organizations like the United Nations perpetuate inequality.

The United Nations is the world's leading development organization and is responsible for setting targets and implementing them in every country in the world. The Sustainable Development Goals were drafted in 2015, to be met in 2030, because it was clear that the very similar Millennium Development Goals (MDGs), aimed at reducing poverty, improving health outcomes and increasing gender equity, established in 2000, would not be met in time. The UN's own reports on MDGs identifies several successes as well as enduring challenges, specifically that the most vulnerable and poorest people, often women and girls, continue to be left behind. The example above indicates the uphill battle against the influence of multinational corporations who seek to profit from inequality, and perpetuate its existence by currying favor with supranational assistance organizations, such as the UN. We argue that political and economic relationships between the UN, multinational corporations and powerful governments have more to do with the problem of producing economic inequality than solving it. **Capitalism** is a global economic system that produces and perpetuates both poverty and inequality and accumulates capital for the already wealthy and privileged (Harvey 1990). We argue that actually meeting the Millennium Development Goals (and the subsequent Sustainable Development Goals) (i.e., reducing inequality) would be troubling for capitalist logic and would undermine, rather than increase, profits. As such, in a system governed by global capital, development follows capitalist logic, and therefore does not produce equity, social welfare or better lives. In this book we aim to illustrate how inequality, of which gender is one form, is composed of many interlocking oppressions perpetuated by **multinational corporations** (firms which operate across state boundaries) and **supranational organizations** (those that operate at the scale of individual country governments).

While this book tackles the enormous problems associated with the gendered impacts of development, it takes its title from the concept of **engendering**, which means "to give rise to." We choose this framing instead of the more traditional "gender and development" because development, like **imperialism**, which is the direct or indirect political-economic control of external territories, and (settler) colonialism, are not past events that have happened and now exist in a historical and apolitical context to study. It continues to happen, and it continues to produce inequality. We emphasize the way "engendering" suggests both an embodied and lived experience that intersects with power, while signaling something new and different from what has come before it in the literature. While gender is in the title, we do not limit our analysis to one form of inequality. We view gender as just one of many intersecting oppressions that are produced and experienced in and through patriarchal and capitalist social relations (Crenshaw 1991).

As such, we take an **intersectional approach** to the study of gender and inequality, which views the ways in which power and subjects are produced at the intersection of many forms of identity: race, class, gender, sexuality, ethnicity, nationality, ability, religion, etc. (Combahee River Collective 1983, Hill-Collins 2000, Mohanty 2005). **Gender** is the socially mediated meaning attached to anatomically sexed bodies, which re/produces ideas of femininity and masculinity and the differences and hierarchies between men and women that are perpetuated by the social relationship of **patriarchy**, or male supremacy (McDowell 2018). **Intersectionality** as a concept emerged from black and queer feminist critiques of white feminisms in the US. Black feminists argue that the **embodied experience** of raced, classed and gendered individuals is shaped in part by an ongoing "matrix of domination" (Hill-Collins, 2000) that contributes to the dehumanization of entire groups of people (McKittrick 2014). We subscribe to the view that everyone experiences varying forms of overlapping oppression and advantage in which we are all complicit (hooks 1997). While we further discuss and define feminism later in the book, we want to signal here that we take a materialist feminist approach, which roots our understandings of the production of gender, difference and inequality in capitalist processes (Mohanty 2005, Hartsock 2011, Hennessy 2017).

As suggested above, there are multiple forms and expressions of feminism. Therefore *feminisms* (plural) rather than *feminism* (singular) more accurately captures this diversity. Throughout this book we will discuss different forms of feminisms such as **liberal-capitalist feminism**, which seeks to empower or emancipate women from patriarchy through economic and political participation within capitalist democracies (Scott 1994). Conversely, **materialist feminism** seeks to add a gendered understanding of the works of Karl Marx and empower women within societal structures associated with socialism or communism (Hartsock 2011). Radical feminists view women's oppression and general inequality as linked to hierarchal systems of governance, and therefore call for the dismantling of unequal systems of governance in order to have gender equality (Mohanty 2005). **Ecofeminists** seek a direct link between the subjugation of women and the environment (Shiva 1991). These various feminist viewpoints have had differential influences on development policies, programs and projects from the 1970s to today.

The diversity of feminist theory also indicates that patriarchal domination is not a single form of oppression that all women (or all men) experience in exactly the same way. Rather, patriarchy is a social relationship defined by one's ability (or lack thereof) to support, perpetuate and benefit from certain kinds of privileged positions that vary across space and time. For example, current systems of economic assistance and development do not regularly incorporate the voices of people in need. Naila Kabeer, a scholar and development consultant, argues that in order to properly address and end poverty, women of color need to be part of economic planning and decision making. However, poverty reduction programs continue to exclude these voices because they are designed by social and economic elites who do not live within spaces of poverty. We take this analysis a step further to assert that economic development requires the creation and perpetuation of intersectional gendered oppressions and inequality, rather than alleviating them (Gibson-Graham 1996). We argue that the modernist notions of social personhood (the bestowal of human rights and citizenship) are denied to most of the world's people under conditions of global capitalism in order to further increase surplus and accumulate capital (Tsing 2015).

We assert that development relies on perpetuating the myth that market forces are mechanisms that can provide public goods and that development will make life better for everyone (Peet and Hartwick 2009). Discourses of development tell comforting stories that obscure how inequalities are produced (Wainwright 2011). For example, why do some people prosper while others do not; and who makes decisions that inform uneven development strategies? Popular notions of development, often in the form of humanitarianism, assume that poor people exist in an ahistorical and far-off place and it is the duty of the wealthy to help them on their way to a better life. These assumptions are rarely questioned; rather they are repeated as part of public discourse. Thus, dominant forms of global capitalism reinforce and support the myth that wealth trickles down to the poor, and corporations are capable of providing public goods to workers, the poor and society in general (Gibson-Graham 1996).

In this book, we take a *post-structuralist intersectional gender analysis of development*. A **post-structuralist** approach means that we view socioeconomic processes as socially produced by human actions (and not externally imposed by structures). Our intersectional gender analyses focus on how development is productive of certain kinds of sex–gender identities, and people whose subject positions as raced, classed, sexed and gendered individuals make them more or less vulnerable to the logic of capitalism. Previous approaches to gender and development addressed the issue of neglecting women from development studies, but often ignored other dimensions of identity (i.e., sexuality, race) or did not critique development as a project and process that has the potential to further entrench rather than alleviate inequality (Benería, Berik and Floro 2003). We argue that an intersectional gender analysis of national and international economic development is important because it pulls back the veil of "goodness" that shrouds much of aid and humanitarianism to reveal how it both functions through capitalism and exacerbates existing inequalities.

We also advocate for **decolonial** approaches to solving the problems of capitalism, which means freeing people from governance that proliferated in the wake of colonialism and that support and perpetuate systems of capitalist accumulation which oppress and dispossess (Tuck and Yang 2012). Mohanty (2005, 7) writes, following Fanon (1963), that decolonization is a process that must happen from the ground up with "transformations of the self, community, and governance . . . through active withdrawal of consent and resistance to domination." We argue that the profoundly negative effects of development on intersectionally and socially constructed subjects everywhere in the world cannot be changed without a systemic reworking of governance and economic systems.

What is development?

Development is a specific socially mediated process through which myriad forms of power work, and through which inequality is produced to generate surplus and capital for the wealthy (Roy 2010). Capitalist economic development takes two general forms: the building and maintenance of an economy, usually a national-scale project, and aid and humanitarianism, usually international in scope and scale. Some mainstream advocates try to distance development from capitalist processes by identifying it as a form of charity that is intended to use economic development as a method of assistance. Geographers have further distinguished big "D" from little "d" development. Big "D" development refers to planned and predetermined interventions with the intention of achieving some type of "progress" or improvement. Little "d" development refers to global capitalist structures that create broad forces or influences of economic change (Hart 2001, Lawson 2007).

Capitalist development exemplifies both forms of development and is favored by many organizations. Capitalism remains integral to North American and

European development organizations, and is premised on the belief that this form of economic structuring will ultimately improve countries and the lives of their citizens. Economic development identifies improvement in various sectors such as agricultural production, healthcare, increased access to technology and mechanization, mitigating poverty, preventing conflict and empowering women. Some projects have indeed helped to improve some people's lives and livelihoods. However, the overarching practices of capitalist development have not adequately addressed power imbalances, the exploitation of places and people, or its effects, which benefit the few at the expense of the many. For example, as middle-class livelihoods grow in some spaces, they are actively eroded in other spaces, and the global division between the world's wealthiest and poorest (countries and people) continues to expand. Additionally, stereotypes about people living in poverty (i.e., as lazy, uneducated, criminal, weak, etc.) continue to dominate capitalist narratives about poverty, thus perpetuating myths about poor people along with incorrectly viewing wealth as aspirational rather than exploitative.

Capitalist development has taken advantage of the vulnerabilities experienced by former colonized spaces by enrolling peasants and working-class people into circuits of capital through the mechanisms of debt and dispossession (Harvey 2003). The machinery of development relies on crushing nonviolent resistance to capitalism through monopolies on coercive force employed by Western countries (mostly former or current imperial powers), and on whose behalf the system of contemporary international economic development has been established, and whose favored allies have benefited enormously (Agnew and Crobridge 2002). In this story of development, we see a clear and necessary intersection with **colonialism**, which is the direct control of people throughout an empire through settlement, land theft, military intervention and/or enslavement. Widely practiced by European empires in the 1400–1800s, colonialism set the stage for capitalism by dispossessing indigenous people of their lands, enslaving workers on multiple continents and using plundered imperial wealth to build an industrial economy for Western elites in the 19th century, also known as **accumulation by dispossession** (Harvey 1990).

The power of Western European nations through colonialism was significantly challenged after World War II. Most European powers concentrated on rebuilding their own countries and economies. Subsequently, colonialism began to fade as independent countries took more dominant roles in managing economies and populations. However, at the same time, a new reorganization of political and economic power occurred through the respective (and oppositional) leadership of the United States and the Soviet Union (also known as the Union of Soviet Socialist Republics – USSR). At the end of World War II, a "Cold War" (based on an arms race) began between the United States and Soviet Union. By 1955, new allegiances were formed and became known as the First, Second and Third Worlds as part of the reorganization of global power. The capitalist economies of Europe and the United States were referred to as the "First World," while the socialist countries of the Soviet Union, China and their respective allies were known as the "Second World." The "Third World," or the "nonaligned" countries, sought a third way or different path for their economic development. Both capitalist and socialist powers sought to influence these countries through economic development programs. Former allies, the United States and Soviet Union, as new "superpowers," competed for economic and political influence throughout the globe. The divergent political-economic ideologies of these powers (capitalism and socialism) sought influence through development projects in various countries.

The Bretton Woods agreement, established in the mid-20th century in the wake of World War II, formed new economic development institutions (i.e., **World Bank**, **International Monetary Fund** (**IMF**)) based on market-based logics of supply and demand (Leyshon and Tickell 1994). These new economic forms required huge inputs of capital, which were out of reach for the poorest countries, many of which were newly independent states. Beginning in the 1980s a new development strategy was created by leaders of the US and UK, known as the **Washington Consensus**. This new strategy sought to manipulate development through granting high interest loans to countries, and market-based reforms that linked development projects to capitalist economic globalization in opposition to Soviet communist-led development.

The Washington Consensus sought to address what was identified at the time as a dependency crisis among countries receiving economic development assistance. The Washington Consensus was so named because it was based on a consortium of ideas from the United States and US-based supranational lending organizations, such as the IMF and World Bank, and trading organizations such as the World Trade Organization (WTO) (headquartered in Geneva, Switzerland). These institutions, along with the US government (in cooperation with other countries, particularly the UK), sought to restructure the global capitalist economy through economic development programs known as **structural adjustment programs** (SAPs). SAPs were designed to provide loans to poorer countries in an effort to address existing economic imbalances. The IMF focused on stability and the World Bank on adjustment methods. These changes also sought internal changes within wealthier countries, leading to the erosion of social assistance and welfare programs, deregulation of corporations and industry, and privatization of government utilities and other government-run companies. Additionally, economic programs focused on eliminating barriers to international trade, leading to the expansion of economic globalization.

Most of the criticisms of the Washington Consensus and SAPs focused on attempts to erase national sovereignty. For example, the growth and influx of nongovernmental organizations (NGOs) in various countries helped to undermine the ability or effectiveness of national governments to attend to the needs of citizens. Other critiques identified SAPs as neocolonial and imperialist, because the benefits of these programs were experienced far more by wealthier and more powerful countries than those on the receiving end of IMF and World Bank loans. Geographers identify the Washington Consensus as a modern form of **financial colonialism**, meaning the beneficiaries of these programs were overwhelmingly wealthy individuals and corporations (Gregory et al. 2011). Additionally, these programs resulted in an extensive increase in debt in countries receiving loans. The Washington Consensus is also criticized for turning the WTO into a "royalty collection agency" for rich countries, due to the benefits experienced by those countries and the growing debt and dispossession experienced by governments and people living in poorer countries (Jahn 2005, 192).

The Cold War ended with the collapse of the Soviet Union in 1991, paving the way for new conceptualizations of capitalist and market-driven economic development. This new paradigm identified economic development assistance as necessary for preventing conflicts throughout the globe, particularly in areas formerly controlled by, or part of, the Soviet Union. Viewing development as conflict prevention intensified after the attacks against the United States on September 11, 2001. The subsequent US-led "War on Terror" further incorporated military violence and security into development projects and programs, particularly in countries such as Afghanistan, Iraq, Somalia, Sudan and Syria. Since the fall of the Soviet Union, China has also become a global leader of economic development. China's method of economic development includes foreign direct investment, resource extraction, promoting China's political position and massive infrastructure plans to link countries in Asia through roads, bridges, railroads and seaports. China's projects in Africa and Asia take a different approach to economic development than its Western predecessors (Bräutigam 2011, Shambaugh 2013, Yeh 2013). Throughout this text we predominantly examine theories and ideologies of intersectionally gendered capitalist economic development championed by Western powers such as the United States and European Union, while also including examples from China. Our examinations focus on the ways in which capitalist development has led to various forms of economic dispossession.

Hartsock (2011) interprets the ongoing enclosure and accumulation in the present day as a continuation of accumulation by dispossession, and adds that it is a process that is profoundly gendered. She writes that accumulation takes place "on the backs of women" and with consequences that intersectionally positioned subjects experience differently. We examine this in a variety of ways in the chapters that follow, from the way in which women resist the seizure of natural resources on indigenous lands, as our cover photo depicts, to the way in which the poverty and inequality that follow dispossession affect men and women differently. Hartsock also demonstrates how the ongoing transformation of labor through globalization contributes to the **feminization of labor**, which is the process of devaluing work and deskilling laborers (both men and women) to further accumulate profits in the private sector. Finally, Hartsock asserts that the enslavement of workers and enclosure of public land and resources are not conditions of the past, but continue into the present in ways that are intersectionally gendered and have consequences, such as forced migrations, land grabs and wage gaps that women and men experience differently.

Throughout this book we use terminology to specifically identify different spatial experiences of development and global capitalism. We refer to the capitalist and formerly imperialist economies of Europe, North America and other neocolonial countries such as South Africa, Saudi Arabia or China as locations for **zones of accumulation**, where capital is hoarded by wealthy elites. These countries frequently extract capital from **spaces of dispossession**, usually territories within current or former colonies such as many African and Asian countries and current Caribbean colonies (and in some cases dispossessing indigenous or racially marginalized people within the boundaries of a wealthy country such as China or the United States).

In between there are mixed economies and/or social democracies, which through policy formations sometimes simultaneously engage in and resist the processes of accumulation and dispossession. We want to highlight that there are wealthy and poor countries but also wealthy and poor individuals within each of these territorial spaces associated with countries. The wealthy often live and work in spaces/zones of accumulation and the economic policies that ensure this wealth rely on spaces/zones of dispossession. As such, people are regularly dispossessed from their access to land, water and resources in order to allow for the unfettered flows and growth of capitalist production and wealth generation for global and local elites. At the same time, local elites in poorer countries benefit disproportionately from development projects compared with their less economically secure contemporaries.

Thus, spaces of dispossession and zones of accumulation can and do exist within every country, regardless of its economic orientation. By using this terminology, we resist the temptation to root places and entire countries within a hierarchical structure, often teleologically presuming a progression from "less developed" to "more developed" status. We also seek to orient our understanding of contemporary development in ongoing capitalist processes that move capital and labor around the world, often to the disadvantage of the poor and vulnerable, further inscribing "differences that make a difference" on entire groups of people. We resist the tendency to refer to countries with cardinal directions ("Western," "Southern"), which we find unhelpfully aggregating, although sometimes we strategically use this terminology as part of historical processes.

Theories and practices of economic development

A country's pathway to development is defined by the structure of its economy as compared with the larger global economy. Independent countries (i.e., not current colonial possessions) can take several approaches to building a national-scale economy. The current favored approach is a **capitalist-export orientation approach**, which emphasizes free trade, open borders and rapid industrialization. These models often start by stimulating agricultural innovations, which include the introduction of agricultural technologies such as the hybrid seeds and chemical fertilizers associated with the **Green Revolution** (discussed further in Chapter 4). In some cases, these innovations increase the quantity of food that is produced on each unit of land, which drives farmland consolidation and displacement of subsistence-based activities undertaken by peasants and newly unemployed landless workers into waged work in manufacturing, often through mass migration to urban centers. This contributes to the feminization of agriculture, as men often migrate and women are left behind in rural spaces to tend plots of land (Radel et al. 2012). Youth, both male and female, also migrate to work in urban centers, and contribute to the largest ongoing migration in history (Saunders 2011).

Waged work relations often require the development of a **service economy**, which is the sector of the economy that provides goods and services, such as haircuts, meals, healthcare and childcare. A service economy emerges alongside the manufacturing sector because wage workers typically migrate without families and no longer have the unwaged work of women providing childcare, food preparation, clothing, etc. Also, in urban areas women are no longer providing basic needs from working the land or within their own households and communities, and are also conscripted into the service sector as wage workers (McDowell 2008). The global capitalist economy is anchored in economic divisions that are driven by the production of industrial goods (cars, clothing, consumer goods) that rely on low waged labor and the consumption of these goods by individuals earning higher wages. Low and middle wage workers are increasingly encouraged to consume at higher rates through access to credit, which for many creates cyclical forms of debt.

Economic development is often measured in terms of productivity (**gross domestic product** (**GDP**)) or income (**gross national product** (**GNP**)), and always in the currency of the most powerful economy in the world. For example, the US dollar is a baseline measure for calculating and comparing different countries' GDP. While economic development is often undertaken by countries within their national borders, the global economy increasingly connects individual countries through sectoral economic activity. The interdependent economic relationship between countries via different economic sectors is often referred to as the **global assembly line** (further discussed in Chapters 2 and 5). The global assembly line is a system of manufacture that sources component parts from multiple countries and regions in the world through a network of flows. These networks, often called **supply chains**, move goods (finished or semi-finished) across space, and are typically administered by multinational corporations, often headquartered in the zones of accumulation.

Globalization is key to this process because capitalists can locate their production facilities in inexpensive labor markets to manufacture, and market to wealthy consuming markets to sell products. Peasants, the poor and dispossessed, vulnerable and marginalized people, are often enrolled as workers in manufacturing these goods at wages well below what their labor contributes to the value of the product (Marx 1867, 1977). For example, Southeast Asian economies are reliant on the manufacture of automobile parts, which are purchased through corporate contracts held by companies in the United States, Asia and Europe. Workers in these factories are migrants, young men and women and other vulnerable workers displaced by conflict or natural disasters and are tightly controlled within their workplaces (Kelly 2002). Criticism of this export-orientation model suggests that it is a predetermined outcome, leading to high mass consumption under conditions of neoliberal capitalism, which is neither achievable by all, nor sustainable given the industrial reliance on nonrenewable resources such as fossil fuels (Banuri 1987, Sparke 2013).

When a newly independent (or newly created) country did not accept the offer of "help" in the form of loans and structural adjustment conditionalities from its former colonizer and/or the colonizer's allies or supranational organizations, that country often tried the route of **self-sufficiency**. Venezuela, Cuba and India, among others, implemented fully or in part self-sufficient economies as a strategy to avoid **dependency** (economic reliance) on former colonizers or the new global institutions of the IMF and World Bank. These strategies usually took a centrally planned approach and focused on invigorating domestic consumption through bureaucratic five-year plans. More centrally planned economies did not rely on mass consumption and accumulation of capital in the private sector as a final goal or an indicator of success, but other measures such as worker health, high employment or various forms of basic welfare. For example, as we discuss below, South Africa provides direct cash transfers to certain groups and has proposed a basic income grant that if approved would provide a minimum income for all citizens (Ferguson 2015).

This approach almost always necessitates isolationist and protectionist trade strategies that make economic growth slow with limited access to goods, while also engaging the majority of workers in productive economic activities. This approach often results in high employment at relatively low wages, entrepreneurial activities and in some cases emphasizes the development of rural places over urban centers. For example, the economy of post-independence India, from 1947 until the early 1990s, was centrally planned using these strategies. Many social economies use democratic decision-making mechanisms for appropriating surplus, reinvesting in the economy and providing for citizens. Under autocratic or authoritarian regimes, however, socialist planning of the economy is used in undemocratic ways, and often enriches elites through the repression of exploited labor. North Korea or Myanmar's military junta regime (1967–2011) exemplifies this kind of centrally planned economy that is undemocratic and dictatorial.

Humanitarianism and aid

Dispossession and debt are often intended consequences of capitalist economic development. In response, the "non-profit industrial" complex (INCITE 2007) steps in, offering a variety of development options under the banner of **humanitarianism**. The geopolitical moments usually preceding capitalist economic development are often colonial or experienced by recently independent countries. Colonialism destroyed indigenous people in order to acquire capital, such as land and accumulated surplus (such as resources) through genocidal forms of dispossession (Coulthard 2014). The loss of land and labor culminated in the decimated capacity of indigenous people to develop or recreate their own pre-existing economies. It also generated dependencies on former colonizers, financial insecurity and/or debts that had to be repaid. It is into this void that humanitarian development steps, offering assistance to the vulnerable colonial/postcolonial subject. After a country or people's land, resources and livelihoods have been stripped away, it becomes much more difficult to refuse development and aid assistance.

A cycle of indebtedness and dependency frequently follows development assistance, which resembles but is not quite the same as colonialism, and is therefore labeled **neocolonialism**. People, state bureaucracies and corporations in zones of accumulation enrich themselves through various forms of legal and quasi-legal land acquisition (i.e., land grabs), low wages, debt slavery, national-scale debt and military interventions abroad and domestically. The economies of those who benefit from dispossession economies appear to ascend the pathway of economic development to achieve high mass consumption (Harvey 1990). Exploitation, discrimination and the construction of inequalities are easily disguised through the rhetoric of **free markets** or those less restricted by government regulation, in capitalist economies of any kind. The rhetorical use of "free" suggests not so much freedom for workers, but the absence of barriers for capital. For example, the words "Work Sets You Free" were inscribed (in German) above the entrance to Nazi concentration camps. Today the use of the word "free" or "freedom" among other popular conceptions (e.g., aid, assistance, development) continues to be manipulated by powerful regimes for their own geopolitical and economic purposes.

The vast majority of development on a global scale today, however, is in the form of export-oriented capitalism that mirrors and mimics colonialism. Many countries have their own development agencies, often located in the zones of accumulation. Governmental agencies use economic development aid or assistance as a method for securing influence, extracting resources or paving the way for market expansion. There are examples of grassroots, democratic development strategies throughout the world, but they are often criminalized, militarily opposed or overthrown by the more powerful (militarily) capitalist countries (Grandin 2011). For example, in 2009 the democratically elected Honduran president Manuel Zelaya was overthrown by the US-backed military for his increasingly left-leaning political leadership in the Latin American region, and his relationships with President Castro in Cuba and President Chavez in Venezuela. Zelaya implemented several socialist reforms in Honduras including providing free education, raising the minimum wage and providing free electricity for the poor.

We assert that development is a form of neo-colonialism designed to enroll governments, natural resources and labor into the circuits of capital so that the very wealthy may continue to benefit from accumulation by dispossession. The socializing of economies, particularly through powerful alliances between countries, is something that threatens the most powerful countries and their supporters. Supranational capitalist authorities, such as the IMF, World Bank and the WTO, construct trade policies and regulate the flow of transnational capital through economic and political manipulation of national-scale governments, who in turn craft policy (in some cases unwillingly) to make labor vulnerable and to exploit natural resources.

Humanitarianism and development are similar: while humanitarianism is generally associated with assisting people to access basic life-giving resources (food, water, shelter) in the aftermath of environmental hazards or political conflict (see Chapter 9), economic development often follows humanitarian interventions, through both big "D" (predetermined) and little "d" (processes of economic change) projects and programs. Nongovernmental humanitarian and development models include organizations such as "Heifer International" and examples of "individualized assistance" where individuals in zones of accumulation such as the United States are asked to donate funds to "give" a cow or other livestock to a poor family in another country. As elaborated in Chapter 3, market-based humanitarian aid/development, in the form of consumer-based giving, provides small forms of assistance, while in some places makes things worse for people, laborers in particular. Micro-credit, what Ananya Roy (2010) calls "poverty capital," is a strategy designed to lift the poor out of poverty, but which has complex and complicating outcomes in terms of the credit and debit cycles within capitalism. The United States Agency for International Development (USAID) provides macro aid in the form of loans and construction – at a cost – usually in the form of multilateral trade or security agreements. Similarly, the World Bank and other supranational organizations provide loans that may help initially in a crisis, but often result in deepening debt and facilitating a capitalist takeover of the economy through the erosion of social safety nets provided by governments such as medical care and social security.

Women (and children) are often positioned as needy recipients of development and aid. Their bodies and lives are an effective tool in "development discourse" to engender pity and consequently action from the individual scale to that of the nation. This is not to say that women and children are not in need in many places; however, the voices of women (and men) in situations of privation should be central to the ideas and economic ideologies of wealthier organizations and governments. The discourse of aid obscures the modes of development, which, as we have already stated frequently, deepen existing inequalities and immiseration. While women, for political-economic reasons, are often (but not always) rendered the most vulnerable subjects of colonialism, imperialism and capitalism, intersectional understandings of identity suggest alternative ways of thinking about women and their relationship to development. We start from the assumption that the categories of women and men are understood relationally through the construct of gender, but that this is just one form of inequality exploited through capitalism, and from which governments and corporations profit.

What is gender?

Anatomical sex differences are given meaning through a binary understanding of men and women as gendered subjects (Butler 1990). Gender identity takes on meaning through ideas and narratives about what it means to be a man (**masculinity**) or a woman (**femininity**). We argue that these categories are relational, meaning that what it means to be a man is defined in opposition to what it means to be a woman (and vice versa) in a given social context, and social expectations are placed on individuals to justify binary gender roles and gendered work cultures, such as the kinds of work that women are thought to be able to do (housework), or to do better than men (electronics manufacturing). For example, antiquated understanding of gender and work associated certain forms of labor, such as cooking and cleaning, with women, and other forms of labor, such as engineering, with men. These gender–work associations are based on social ideologies rather than biological differences between men and women. There is no biological indicator that clearly delineates male or female proclivity or ability to conduct certain types of work.

Conservative ideas about gender and sex, usually defined by a resistance to change, reinforce binary gender roles, and this remains the norm in many institutions. This persists in spite of the prevalence of intersex individuals who are both or neither anatomically sexed at birth, and as a result of binary ideas about sex and gender, intersexed children are medically altered as one sex (either male or female). Challenges to binary gender roles can be seen among the growing number of people who publicly assert a spectrum of gender identities. Among them is **transgender**, or the identity of people who transition their corporeal (i.e., bodily) appearance in some way at some point in their lives to accurately reflect the gender with which they identify. These new categories of gender rework historically binary notions, and include the identity **cisgender**, which is the label used to identify people whose gender, or social identity, corresponds to the sex they were born into or were identified as belonging to at birth. There are also **nonbinary gender** identifying people, who do not wish to subscribe to or be identified as fitting within an either/or gender binary.

Issues of gender are clearly related to notions of sex and **sexuality**, or the meanings and practices attached to desire and reproduction. Judith Butler (1990), a gender theorist, argues that gender is a conditional performance and the product of compulsory heterosexuality. Gender is not simply an identity linked to anatomical sex – it is a socially constructed set of performances that are also conditioned by the cultural economies of work (Hanson and Pratt 2003), and **heteronormativity** (the idea that heterosexual is or should be the norm). Butler (1990) argues that notions of gender matter under conditions of modernity because we are enmeshed in the social relations of the "**heterosexual matrix**," which is often enforced through law. For example, India decriminalized consensual sexual relations between same-sex couples in 2018, the same year that a same-sex Malaysian couple were publicly caned for engaging in consensual sexual relations. Butler argues that people are encouraged or coerced into performing a given gender and sexuality to achieve social personhood through the institutions of family, religion and civil society.

Sex is commonly thought of as strictly related to anatomical differences, while **gender** consists of social meanings attached to anatomical bodies. Through repeated performances, we see the emergence of meanings around what it means to be a man (masculinity) and a woman (femininity) that correlate to anatomically sexed bodies. We can also observe that in some contexts third genders or nonbinary identities disrupt these notions. Butler, however, argues that both sex *and* gender have "performances" attached to them that reinscribe differences that "make a difference." In other words, society makes sex differences meaningful, when they may not be important to anything or anyone. She argues that **compulsory heterosexuality** promoted by religions and governments for the purposes of producing and reproducing a population of workers and those who will perpetuate cultural value systems (e.g., religious adherents) drives the significance attached to sex and gender. This emphasis results in a set of behavioral expectations for women and men that may have little to do with how their bodies look, or what they can do.

Because we take an intersectional approach to gender in this volume, our analysis of the impact on development on men and women does not begin or end with examining sex, gender and sexuality. Rather, we include other axes of identity, such as **race**, the socially imposed category of identity that is thought to correspond to melanin content in the skin, and which places individuals in social hierarchies accordingly; and **class**, the social category most closely associated with income, but also placement in social hierarchies through birth, inheritance and other forms of identity. We also examine gender through the lens of **nationality** (country of origin or citizenship); **ethnicity** (social categories associated with family, tribe, religion and nation) and **caste** (social categories often associated with occupation, and into which people are born). We examine identities in relationship to capitalism as shaped and inflected by other social categories. In order to understand any particular person's relationship to development, it is imperative to also examine these relations within the context of other social dynamics.

Gender and development: a history

Including women as part of development theories and policies began in the 1970s, influenced by Ester Boserup's (1965) book, *Women's role in economic development*. This book identified the various ways in which women were earning wages and actively involved in agricultural production. Both colonists and later development professionals viewed women's labor as *only* existing inside the home and therefore did not recognize women's contributions, particularly to agricultural production (Wilson 2015). The United Nations responded to these findings by identifying 1975–85 as "The United Nations Decade for Women," beginning with the International Women's Year Conference in Mexico City. The United Nations Fund for Women (UNIFEM, now called UN Women) was also established at that time. A focus on Women in Development (WID) took a liberal modernization approach, which included a restructuring of development programs to include women and focused on welfare, equity, efficiency, empowerment and ending poverty (Peet and Hartwick 2015, 284). The WID approach was mostly adopted by formal governance institutions and government-backed development agencies. It has also been critiqued for falling short of its intended goals of assisting women. For example, in 1985 at the end of the Decade for Women, limited improvements were accomplished and many women experienced less access to resources, increased work burdens along with poor health and nutrition.

Women participants at the UN-sponsored conferences in 1980 (Copenhagen) and 1985 (Nairobi) intensely debated feminist theories, imperialism, racism and national identities. Women located in the Third World advocated for incorporating gender with their intersectional identities (based on race, ethnicity, class, etc.) and other participants discussed their opposition to capitalism. Thus, women called for alternative development approaches and socialist development, which led to the Women and Development (WAD) approach that focused on challenging global capitalism and patriarchy. WAD advocates identified modernization and capitalism as integral to women's oppression. Marxist-feminist approaches to socialist development were central to WAD perspectives (Peet and Hartwick 2015). WAD identifies women's productive and reproductive labor as equally important for functioning economies. "In general the idea was that women performed unpaid labor in reproducing labor power as a kind of subsidy for capital as well as working directly for capital as employees in factories or producers of commodities" (Peet and Hartwick 2015, 291). Socialist feminists also believed it was necessary to transform gender relations from unequal to egalitarian in order to have effective development.

Gender and Development (GAD) was another perspective that also originated in the 1970s. GAD focused on gendered divisions of labor in order to understand how dependency works between men and women. In this way, it differed from WID, which did not directly question gender divisions of labor, but rather sought to assign equal value to men's and women's work. Thus, for GAD, gender relations, rather than "women," became the primary category for analysis and program design. In the 1980s, women from spaces identified as the Third World and "developing" countries began advocating for programs that sought to empower women as agents of change rather than continue with WID and WAD programs, which identified poor women as problems needing to be solved (Peet and Hartwick 2015). These critiques culminated in the formation of Development Alternatives with Women for a New Era (DAWN), which was established in 1984 in Bangalore, India.

The DAWN platform called for a comprehensive and complex understanding of history and disparate geographies in order to counter capitalist economic development, which relied on powerful patriarchal hierarchies and the subjugation of women (Kabeer 1994). DAWN also made significant contributions to GAD theories and approaches to development. Other gender and development theories focused on environmental issues. Women, Environment and Development (WED), also originating in the 1970s, identified men's control over nature as parallel to men's control over women. WED development theories fit well with ecofeminists who similarly connected the subjugation of women with degradation of the environment. For example, ecofeminist activists – such as Vandana Shiva, founder of Navdanya "nine seeds" – developed women-led and women-focused social movements to protect the environment and biological diversity (Mollett 2018).

For example, the 1960s international agricultural development referred to as the **Green Revolution** in India exacerbated existing inequalities to the point of fomenting **communal** (differences associated with caste and religion) violence and genocide, by exhausting scarce water resources, consolidating farmland into the hands of the already wealthy and displacing workers, peasants and the poor, typically lower caste and from the least favored religions, from the land (Shiva 1991). Additionally, the production of pesticides as part of the Green Revolution in India included one of the worst industrially driven environmental disasters in the country. In 1984, the Union Carbide chemical company's Bhopal facility experienced a leak of pesticides, which exposed over 500,000 people to toxic chemicals. In the immediate aftermath of the leak, 2,259 were killed from exposure, and in subsequent days the death toll rose to 3,787. The leak is also responsible for 558,125 injuries and over 3,000 permanently disabled persons. Civil and criminal cases involving Warren Anderson, the CEO of Union Carbide, and the company were initiated in India, leading to small fines. However, clean-up efforts in Bhopal were not adequately conducted by the company or the government of India, leaving the people of Bhopal with contaminated water and ground soil (Hanna, Morehouse and Sarangi 2005).

While the Green Revolution had an impact on women's work, it also deeply affected individual men as well as influenced generally held notions of masculinity. It is widely recognized in recent years that men and masculinity are missing pieces to the study of gender and development, as well as the interventions that such an analysis might inspire, such as the masculinity of militarism, or the paternalism of philanthropy (Kimmel 2002). However, little has been done to include men (particularly men in zones of accumulation) into an analysis and discussion of development and its impacts. This could include, for example, research and inquiry into the relationship between hegemonic masculinity, capitalism and the production of inequality. Additionally, and relatedly, few approaches to understanding development and its relationship to embodied identity begin from a position of intersectionality (Harcourt 2016). An intersectional approach begins with the assumption that there are diverse kinds of women-identified people who encounter development in many divergent ways in different kinds of places. It further investigates how the object of development specifically constructs an "other" produced through the racist and sexist legacies and aftermaths of colonialism and imperialism (Rizzo and Gerontakis 2016). Such an approach asks important questions about the production of difference that makes a difference through the technologies of the heterosexual matrix, the role of territory and capital, and how social hierarchies emerge as the "displacement of difference" (McKittrick 2006).

We view development as capitalism structured in such a way that intersectional differences (of which gender is just one) are embedded and enrolled in circuits of capital accumulation in place-specific ways. First, we argue that in order to ensure "better lives for all" we must first take an intersectional approach to understand why some have better quality of life than others, and to understand what constitutes quality of life in the first place. Gender and development studies recognize that when women and women's lives are considered, development processes and outcomes are also different (Benería, Berik and Floro 2003). We argue that considering how intersectional identities are produced spatially, at a variety of scales in place-specific ways, and how that impacts development processes and outcomes, is a vital next step. For example, conditions that may create vulnerability for one group in one place may not create them in the same group in another. Secondly, we assert that capitalism is dubious as a framework for engendering a better quality of life for any single person, much less all people. Governments are not always helpful allies because neoliberal capitalism requires countries to facilitate rather than regulate the private accumulation of capital (Trauger 2017). Last, we assert that alternatives are possible, which emerge autonomously in place-specific ways through democratic rather than neo-imperial mechanisms. Geographical context is thus key to understanding the relationship between power, identity and the economy, from the scale of the body to the global economy, and how to work for transformations that work for all people.

Chapters and themes in the book

There are three central themes to which we will return throughout the text: 1) development exploits and exacerbates existing inequalities and produces new ones; 2) gender is relational and intersectional; and 3) recipients of aid/investment are socially constructed as "needing" help in the form of development. We use these themes to discuss how capitalist development (via imperialism) produces intersectional forms of difference, which are then used to reproduce and perpetuate certain kinds of unequal economic and social arrangements within places and across space. **Intersectional gender**, or gender that considers the influence of other forms of identity, is one form of difference (among many) that is used to render people vulnerable to exploitation, which reinforces the power and influence of those who already benefit from capitalism. Forms of identity are further inflected with power relations in an intersectional analysis. For example, a white woman may experience discrimination based on her sex but not on her race, while a black woman (especially in the United States) can experience discrimination based both on her gender (female) and race (black). Aid and development, while purporting to help those who are least empowered, masks the way in which capitalism always produces inequality along these axes of differences (i.e., white women are paid less than white men, and black women are paid less than white women), and disempowers and disenfranchises the most vulnerable people. All of these processes play out within and across place and space, and effective solutions are place-specific and not one-size-fits-all.

In the next two chapters, we expand on the intellectual engagement with the concepts and theories of gender and development in Chapter 1. We discuss how intersectional gender, class and racial hierarchies emerged as relational categories. These divisions served as an organizing principle in the political-economic histories of colonialism and imperialism and under conditions of capitalist development today. We aim to connect the history of Western European expansion and US colonialism to discuss how development emerges, is embraced and contested at multiple scales. In Chapter 3 we examine the business of aid and development and demonstrate how celebrities, for-profit businesses, nonprofit organizations and countries use development as a social, political and economic tool to generate or secure wealth, power and sociopolitical influence. Development-as-capitalism produces subjects in need of aid, which businesses, consumers, celebrities and nonprofits attempt to "rescue"; however, philanthropy works as a form of capitalism to generate capital surplus from desperation and dispossession. Chapter 3 explicates how advertising narrowly constructs those in need (mostly as women and children without agency, capacity or ideas), while simultaneously representing wealth and the wealthy as aspirational and philanthropic rather than the perpetrators of inequality and poverty.

Chapters 4 through 6 outline how economic activity is a necessary precursor to the development of an independent country. We examine three sectors of the economy to understand the way intersectional identity is used as an organizing principle in the process of development. We begin by examining how raced, classed and gendered subjects, such as farmers and laborers, are exploited by countries and the private sector to accumulate capital, often through dispossession in the global economy. We argue that the mechanisms of development outlined in Chapter 4 make certain groups particularly vulnerable to exploitation. In Chapter 5 we examine how capital investments, in the form of land acquisition, infrastructure projects and housing markets, serve as pathways to development through the exploitation of gendered, raced and classed subjects. We outline in Chapter 6 how and where manufacturing, service and knowledge economies develop and how they enroll individuals in circuits of capital. The scholarly contributions of geographers inform our understanding of gendered work in the areas of manufacturing, service industries and knowledge economies.

Chapters 7 through 9 shift focus from dynamics to what we identify as critical moments in development. These moments require place-specific understandings to generate relevant and responsive outcomes in the realms of both policy and practice. In Chapter 7 we discuss the relationship between development and the gendered dimensions of health, healthcare and population management. We examine the gendered links between public health and population management/control, and the gendering of public health and healthcare measures as development imperatives. Chapter 8 takes up the issues of technology transfer as part of development practice. While technology is considered integral to development and as a method for fixing economic problems, it is not value-neutral. In this chapter, we focus on the production of subjects "in need" of technology, and the way technology works to create and protect capitalist markets. In Chapter 9 we examine aid and development in the aftermath of environmental disasters and political conflict. Conflict development includes the provisioning of funds for private security, military and logistic firms. Humanitarian assistance in the wake of disaster or conflict brings in needed resources and supplies, and attends to vulnerable individuals. Simultaneously, some individuals working for humanitarian organizations take advantage of human vulnerability and their position of power by trafficking humans, and sexually abusing or harassing individuals.

In Chapter 10 we conclude by examining the emerging new political possibilities for post-development and decolonial alternatives. These include social movements to remake democracy, radical organizations to resist occupation and dispossession and efforts to challenge and change economic forms that oppress and exploit. We examine transnational feminist organizations, global social movements and alternative political economies to present different methods for making a living and coping with social, economic and environmental change that supports those who are the most vulnerable, not the most powerful. We conclude by summarizing various theories and ideas from post-development scholarship, and encouraging readers to conceptualize the global economy in ways that benefit the least powerful people.

Recommended reading

A small place, Jamaica Kincaid; *Sisterhood: political solidarity between women*, bell hooks; *Dead aid*, Dombisa Moyo

Recommended viewing

Babies; *Life and debt*; *End of poverty?*; *No logo*; *Dam/aged*

Questions for discussion

How do you see economic development or aid and humanitarianism in your everyday life? What influences your identity as an intersectionally gendered subject? What would cities look like if they were designed for mothers?

References

Agnew, J., & Crobridge, S. (2002). *Mastering space: hegemony, territory and international political economy*. London: Routledge.

Banuri, T. (1987). *Modernization and its discontents: a perspective from the sociology of knowledge*. World Institute for Development Economics Research. University of Massachusetts.

Benería, L., Berik, G., & Floro, M. (2003). *Gender, development and globalization: economics as if all people mattered*. New York and London: Routledge.

Bräutigam, D. (2011). "Chinese development aid in Africa." In J. Golley, & L. Song (Eds) *Rising China: global challenges and opportunities*, Canberra, Australia: ANU Press (pp. 203–22).

Butler, J. (1990). *Gender trouble: feminism and the subversion of identity*. New York, NY: Routledge.

Combahee River Collective (1983). *Combahee River Collective statement: black feminist organizations in the 70s and 80s*. New York, NY: Kitchen Table/Women of Color.

Coulthard, G.S. (2014). *Red skins, white masks*. Minneapolis, MN: University of Minnesota Press.
Crenshaw, K. (1991). Mapping the margins: identity politics, intersectionality, and violence against women. *Stanford Law Review*, 43(6), 1241–99.
Dottridge, M. (2002). Trafficking in children in West and Central Africa. *Gender & Development*, 10(1), 38–42.
Fanon, F. (1963). *The wretched of the earth*. New York, NY: Grove Press.
Ferguson, J. (2015) *Give a man a fish: reflections on the new politics of distribution*. Durham, NC: Duke University Press.
Gharib, M. (2017). World leaders gobble up M&M's imprinted with U.N. goals. www.npr.org/sections/goatsandsoda/2017/09/22/552728625/world-leaders-gobble-up-m-ms-imprinted-with-u-n-goals. Accessed 10/23/2017.
Gibson-Graham, J.K. (1996). *The end of capitalism (as we knew it): a feminist critique of political economy*. Oxford: Blackwell.
Grandin, G. (2011). *The last colonial massacre: Latin America in the Cold War*. Chicago, IL: University of Chicago Press.
Gregory, D., Johnston, R., Pratt, G., Watts, M., & Whatmore, S. (Eds) (2011). *The dictionary of human geography*. Oxford: John Wiley & Sons.
Hanna, B., Morehouse, W., & Sarangi, S. (Eds) (2005). *The Bhopal reader: remembering twenty years of the world's worst industrial disaster*. Goa, India and New York, USA: The Apex Press.
Hanson, S., & Pratt, G. (2003). *Gender, work and space*. London: Routledge.
Harcourt, W. (Ed.) (2016). *The Palgrave handbook of gender and development: critical engagements in feminist theory and practice*. London: Springer.
Hart, G. (2001). Development critiques in the 1990s: culs de sac and promising paths. *Progress in Human Geography*, 25(4), 649–658.
Hartsock, N. (2011). A new moment of primitive accumulation. In Conferencia Inaugural de la Conferencia Inkrit. http://inkrit.de/mediadaten/pdf/inkrit11hartsock.pdf.
Harvey, D. (1990). *The condition of postmodernity: an enquiry into the conditions of cultural change*. Oxford: Blackwell.
Harvey, D. (2003). *The new imperialism*. Oxford: Oxford University Press.
Hennessy, R. (2017). *Profit and pleasure: sexual identities in late capitalism*. London: Routledge.
Hill-Collins, P. (2000). *Black feminist thought: knowledge, consciousness, and the politics of empowerment*. New York and London: Routledge.
hooks, b. (1997). Sisterhood: political solidarity between women. *Cultural Politics*, 11, 396–414.
INCITE (2007). *The revolution will not be funded: beyond the non-profit industrial complex*. Durham, NC: Duke University Press.
Jahn, B. (2005). Kant, Mill, and illiberal legacies in international affairs. *International Organization*, 59(1), 177–207.
Kabeer, N. (1994). *Reversed realities: gender hierarchies in development thought*. London: Verso.
Kelly, P.F. (2002). Spaces of labour control: comparative perspectives from Southeast Asia. *Transactions of the Institute of British Geographers*, 27(4), 395–411.
Kimmel, M.S. (2002). *Masculinities matter! Men, gender and development*. New York, NY: Zed Books.
Lawson, V. (2007). *Making development geography*. London: Hodder Arnold.
Leyshon, A., & Tickell, A. (1994). Money order? The discursive construction of Bretton Woods and the making and breaking of regulatory space. *Environment and Planning A*, 26(12), 1861–90.
Marx, K. (1867, 1977). *Capital, vol. 1*, trans. Ben Fowkes. New York, NY: Vintage.
McDowell, L. (2008). Thinking through work: complex inequalities, constructions of difference and trans-national migrants. *Progress in Human Geography*, 32(4), 491–507.
McDowell, L. (2018). *Gender, identity and place: understanding feminist geographies*. London: John Wiley & Sons.
McKittrick, K. (2006). *Demonic grounds: black women and the cartographies of struggle*. Minneapolis, MN: University of Minnesota Press.
McKittrick, K. (Ed.) (2014). *Sylvia Wynter: on being human as praxis*. Durham, NC: Duke University Press.
Mohanty, C.T. (2005). *Feminism without borders: decolonizing theory, practicing solidarity*. Chapel Hill, NC: Duke University Press.
Mollett, S. (2018). "Environmental struggles are feminist struggles: feminist political ecology as development critique." In A. Oberhauser, J.L. Fluri, R. Whitson, & S. Mollett (Eds) *Feminist spaces: gender and geography in a global context*, New York, NY: Routledge (pp. 155–187).
Peet, R., & Hartwick, E. (2015). *Theories of development: contentions, arguments, alternatives*. New York, NY: The Guilford Press.
Radel, C., Schmook, B., McEvoy, J., Mendez, C., & Petrzelka, P. (2012). Labour migration and gendered agricultural relations: the feminization of agriculture in the ejidal sector of Calakmul, Mexico. *Journal of Agrarian Change*, 12(1), 98–119.
Rizzo, T., & Gerontakis, S. (2016). *Intimate empires: body, race, and gender in the modern world*. Oxford: Oxford University Press.
Roy, A. (2010). *Poverty capital: microfinance and the making of development*. New York, NY: Routledge.
Sandoval, C. (2000). *Methodology of the oppressed*. Minneapolis, MN: University of Minnesota Press.
Saunders, D. (2011). *Arrival city: how the largest migration in history is reshaping our world*. New York, NY: Vintage.

Scott, J.W. (1994). "Deconstructing equality-versus-difference: or, the uses of poststructuralist theory for feminism." In S. Seidman (Ed.) *The postmodern turn: new perspectives on social theory*, Cambridge: Cambridge University Press (p. 282).

Shambaugh, D.L. (2013). *China goes global: the partial power*. Oxford: Oxford University Press.

Shiva, V. (1991). *The violence of the green revolution: ecological degradation and political conflict*. London: Zed Books.

Sparke, M. (2013). Introducing globalization: ties, tensions, and uneven integration. West Sussex: Wiley-Blackwell.

Trauger, A. (2017). *We want land to live: making political space for food sovereignty*. Athens, GA: University of Georgia Press.

Tsing, A. (2015). Salvage accumulation, or the structural effects of capitalist generativity. Cultural Anthropology website, March 30. https://culanth.org/fieldsights/656-salvage-accumulation-or-the-structural-effects-of-capitalist-generativity.

Tuck, E., & Yang, K.W. (2012). Decolonization is not a metaphor. *Decolonization: Indigeneity, Education & Society*, 1(1), 1–40.

Wainwright, J. (2011). *Decolonizing development: colonial power and the Maya*. John London: Wiley & Sons.

Wilson, K. (2015). Towards a radical re-appropriation: gender, development and neoliberal feminism. *Development and Change*, 46(4), 803–32.

Yeh, E.T. (2013). *Taming Tibet: landscape transformation and the gift of Chinese development*. Ithaca, NY: Cornell University Press.

2 Engendering development

Introduction

It is 1619. A slave ship departs Accra, Ghana. It is filled with men and women destined to work in British colonies on the sugar plantations of the Caribbean. They were prisoners of war or the spoils of inter-ethnic raiding to be bought and sold until they were purchased by Portuguese slave traders. The people who purchased them viewed them as subhuman and only worth caring for in terms of their economic productivity. The men were forced to cut cane, and the women to provide for the men and to reproduce the next generation of slaves, often through systematic rape by slave owners. The space of the ship contained not only the real human capital of empire, but also the powerful knowledge systems that justified slavery through racial categorization, as well as the production of gendered work and identity along lines of sexual difference. The racial-sexual violence of empire that is embedded in the economic transactions produced the capital that built the contemporary economic systems and forged many of its social relationships (McKittrick 2006).

As discussed in Chapter 1, masculinity and femininity are relational constructs, developed in concert with other forms of hierarchical differences, such as race, class and ability. These categories emerge through the process of economic development as a way to subordinate and exploit certain groups, such as slaves, for profit. Surplus is generated through paying workers less than their labor is worth, and hierarchical social systems (informed by patriarchy and white supremacy) identify and justify those who will be paid less, or not paid at all, in the interests of capital accumulation and profit. In what follows we describe the advent of contemporary capitalism and demonstrate how hierarchical thinking in the age of Enlightenment and within the spaces of Western European empires were central to the creation of knowledge and belief systems that remain part of contemporary economic development projects. In this chapter we provide a brief history of empire and how imperial ideologies produced identity categories that are central to development today.

A short history of empire

Empires and imperialism emerged as a dominant form of social organization in human societies several thousand years ago. The most recent empires – which declined in the 20th century – gave way to new types of social organization – the nation-state, as well as new forms of imperialism. The most influential in terms of what we now understand as economic development in the modern era were the British (and other European) empires, the USSR as well as the American Empire. These disparate groups have three similar principles of imperialism: the projection of power across space, the acquisition of territory and the use of settler colonies. **Imperialism** can be thought of as an ideology of dominance for the purposes of obtaining territory or

FOCUS: THINKING RELATIONALLY

Categories take on meaning through their relationship to other categories. Often referred to as binary thinking, one category, such as "rural," becomes defined by what it is not: "urban." Feminist scholars have long critiqued this way of thinking, because it is inherently value-laden, and arbitrarily presumes that categories on one side of the binary are less desirable than the other. Concepts arising from Western European Enlightenment, such as rationality and reason, were associated with men, while women were viewed as irrational and emotional. Feminist and queer scholars have analyzed this way of thinking to understand how categories take on meaning through discourse and thinking relationally (Goldner 1991). For example, relational thinking helps expose the values and ideologies that support femininity and masculinity as binary categories. Exposing this pattern creates an opening to think about gender identities as fluid and overlapping, rather than fixed and static categories. It also raises awareness about intersectional identity formation, or when identity takes on multiple meanings through the accretion of many kinds of difference.

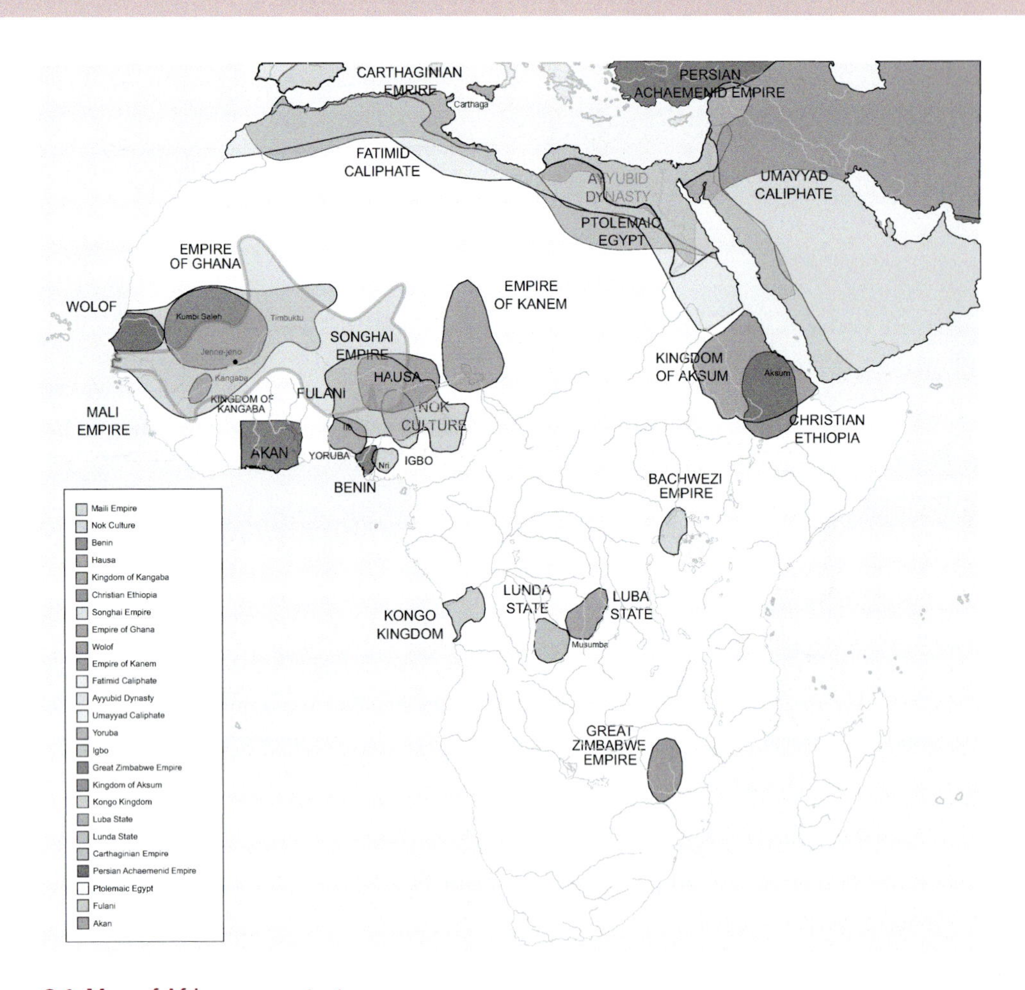

Figure 2.1 Map of Africa pre-contact

Source: by Jeff Israel (ZyMOS) [GFDL (www.gnu.org/copyleft/fdl.html) or CC BY-SA 3.0 (https://creativecommons.org/licenses/by-sa/3.0)], from Wikimedia Commons

FOCUS: ORIENTALISM AND INTERSECTIONALITY

The negative associations with certain phenotypes is also related to **orientalism**. Orientalism, coined by Edward Said (1977), is a way of thinking about people in the world that emerged at the same time European empires were embarking on colonial projects in places throughout the world, including East and South Asia. The values of the Enlightenment that supported ethnic superiority and racial hierarchies reified ideas that European colonizers were superior to the colonized "other," even and especially when confronted with evidence that colonized thinkers and writers exceeded European scholarship and technological prowess. Ignoring women's contributions to this work, because in their worldview women were not capable of reason, colonizers focused on denigrating men. Framed as feminine, incapable of governing, or as hyper-sexualized and violent, colonizers justified the imprisonment, execution, re-education and subjugation of colonized peoples. The impact of these discursive frames continues into the present day, and can be seen in entertainment, media and the persistent over-policing of nonwhite bodies in European and American contexts. See also Reina Lewis (2013) for a gendered perspective on orientalism.

influencing power relations at a distance. **Colonialism** is a practice of empire, which manifests through a multitude of forms, such as military interventions, religious missionars, coups and infrastructure development. The point of imperialism and the colonization of territory and people is to accumulate capital and labor for the interests of the most powerful.

Our story begins with what is known as the "Enlightenment," which ushered in the era of **modernity** (from the mid-1400s and early 1500s to the late 1900s). This is a period characterized by imperial acquisition of territory outside of the European continent and the development of science, political philosophies of freedom and progress and social hierarchies based on bodily appearances, also known as a **phenotype**. The apparent liberalism of the Enlightenment ideas, which supported freedom and liberty for all, were at apparent odds with the increasing acquisition of slaves by Europeans and the genocide of indigenous people on the American and Australian continents. This indicates the degree to which social hierarchies allow the incomplete transfer of rights to all people, even in the context of liberating rhetoric, and indicates how racial categories support and perpetuate the ongoing privilege of elite classes and some races of people (Bouie 2018).

The acquisition of territory by European empires left few spaces on the face of the earth untouched. By 1914, nearly 85 percent of the world's population lived in areas controlled by Europeans (Rizzo and Gerontakis 2016). Modernist ideas about "rational thought" justified much of this control. Historically, modernity is thought to begin with the work of Machiavelli, who rejected medieval rationalities focused on fatalism and tradition, in favor of normative notions of what could be possible for society. Other Enlightenment thinkers (Hobbes, Kant, Locke, Smith) expanded on these ideas to develop republics composed of rights-bearing citizens (as opposed to monarchies composed of serfs and subjects). Additional examples include mercantile capitalism and applying natural science to order society, which is also known as **eugenics** or **social biology** (see Chapter 7). Of primary importance to imperial practices was the distorted use of science and notions of progress to justify the acquisition of new territory (i.e., in Africa, Asia, Oceania and South America) to bring what the Marquis de Condorcet, who lived in the mid-1700s, called "freedom," "civilization" and "reason" to people living outside of Europe. In practice, this resulted in the subjugation of people based on phenotypical characteristics such as head circumference and skin color. Phenotypical characteristics were classified, following Darwin's work on the origin of species (or races, as they were called), to develop what are now identified as racial categories.

McKittrick (2014) argues that the orientalism and eugenics of the Enlightenment period brought about the emergence of categories of "nonhuman" and "human" based on observable traits. This would evolve into the modernist relational understanding of personhood based on categories of race, dis/ability, gender, religion and class. In the modern era, humanness came to be defined by whiteness, ability, heterosexuality, Christianity and capitalist (land-owning) maleness. "Nonhuman," according to McKittrick (2014), was, and continues to be, based on intersecting combinations of race, class, gender, etc., and generated hierarchical knowledge systems. These racial categorizations of people were considered as undeserving of personhood. Many of these racist assumptions persist today, and are visible in police brutality against young men of color in the United States, the rapes and murders of indigenous women by white men on reserves in Canada and the disappearance of women who organize to resist corporate domination (Wright 2006). McKittrick (2014, 143) writes that

> social-spatial expressions of Western modernity – colonial encounters during and after the fifteenth and sixteenth centuries . . . territorial expansion and transatlantic slavery, industrialization, the rise of biological sciences – accumulated and formed overlapping governing codes (Man1 and Man2) as overrepresentations of the human.

In this typology, based on the work of Sylvia Wynter, McKittrick identifies Man1 as the imperial, theological subject of the pre-Renaissance era, and Man2 as the category of humanness that emerged from the Enlightenment, which was constructed in opposition to black, indigenous and female persons, and continues today through a logic of capitalist accumulation. McKittrick asserts that these categories are "inflected by powerful knowledge systems and origin stories that explain who/what we are" (2014, 10). The people who fell into racial categories that were associated with humanness benefited most from this system and laid the groundwork for social categories that persist today.

Although sexual identities are thought to be innate, our contemporary understandings of what constitutes sex and gender are based on histories, myths, discourses and practices that relate to anatomical difference. Norms of masculinity emerged in various ways in different cultures, but European masculinity was defined in relation to imperial practice. Rationality was based on scientific reason; ability was based on military conquest and the development of mercantilist capitalism separated men's work from women's work. These values merged to form a particular kind of masculinity in sharp contrast to the colonized other, who was identified as weak, subordinate and effeminate, or violent and in need of control. Social Darwinism (later known as eugenics and social biology) was used to introduce the idea of "race" through warped biological understandings of speciation and survival. These notions were applied to entire non-

FOCUS: MODERN SLAVERY

There are more enslaved people today than in the entire history of slavery. Estimates indicate that more than 30 million people are currently enslaved. Modern slavery exists in a variety of industries, particularly agriculture and food production, sex work, domestic work, mining and construction. People are enslaved when they are trafficked illegally into work outside their home countries and are held against their will and forced to work. Debt slavery is also a common mechanism of forcing work without payment. Debt bondage can occur when a person is smuggled over a border and cannot repay the smuggling fee, or when workers borrow money from employers but do not earn enough to pay back the loan. Sex workers, particularly children, are frequently forced to work against their will through a variety of mechanisms including debt, trafficking and forced marriage (Bales 2012).

European societies and were used to cast Europeans as more advanced and civilized than other races. For example, Europeans valued wide social and spatial divisions between men and women, and saw them as an indication of advanced civilization (as opposed to the unequal status of women). Private spaces, such as homes/domestic sites, were associated with women/femininity and public spaces such as offices, factories and politics were associated with men/masculinity. On this basis, they justified the introduction of sex segregation in societies where space was not divided by gender (Domosh and Seager 2001).

In feudal European societies prior to the age of Enlightenment, men and women often worked together on the same kinds of agricultural tasks in village contexts. As European empires grew domestically and internationally, new forms of work, as well as new urban centers, emerged. The extraction of raw materials (natural resources, agricultural products) from the colonies, as well as the privatization of resources within the **metropole**, or imperial countries, led to the development of manufacturing. For example, cotton grown in the milder climates of British colonies in India, North America and Egypt led to fabric manufacturing in northern England, to which men migrated to work for wages in factories (also known as **productive work**). Women and children accompanied them to the city, leaving behind less and less viable agricultural labor, to work in the domestic sphere of the home to provide for workers. The division of labor into factory and domestic work led to ideas about femininity (domesticity, passivity) that were exported to the colonies (Domosh and Seager 2001). Some European women traveled to the colonies as adventurers, scientists, wives, nurses, teachers and missionaries, but very few escaped the ideas of femininity that restricted the kind of work they could do (and with whom), as well as their mobility.

The emergence of racial–sexual identities in the age of imperialism is apparent in the regulation of sexual relations between settlers and indigenous people. Sex between colonists and natives took a variety of forms, and included temporary or permanent consensual as well as nonconsensual relationships between white men and native men and women. European men stood to gain access to culture and language through a sexual partner, and were often living in places without laws against rape, or the more restrictive traditions of their homelands. Mixed race children from these unions were sometimes granted certain privileges of whiteness based on arbitrary (and false) ideas about **blood quantum** (or percentage of racial heritage inherited from parents). Settler women were not granted the same sexual privileges as men; many did, however, find ways to have sexual relationships with indigenous men. Consensual sex between white women and indigenous men was viewed as impossible because of the entrenched taboos. White men constructed white women as pure, and therefore any sexual relationship with nonwhite men was seen as a defilement of that purity. White men used these entrenched conceptualizations of white women to justify the murder of black or indigenous men well into the 21st century (Faludi 2007). **Miscegenation** laws – those that forbid interracial marriage – were only struck down in the United States in 1967, and some countries, such as India, continue to influence intimate relationships between disparate groups (Smith 2012).

Social elites among the colonial populations were also significant in shaping knowledge about the colony. Most pre-contact societies had developed and expanded their own systems of science, education and governance prior to colonization. For example, Indian republics existed long before British colonialism, and developed sophisticated scientific technology, such as the Jantar Mantar astronomical observatory in Delhi. The Ottoman Empire produced important navigational technologies such as the sextant, and trained women to be physicians well before European women. Ottoman institutions were diminished in significance, and in many cases evidence of their achievements were erased, as part of the process of imperialism by the British. Throughout their colonies, the British created or reinforced social and political elites as their local allies. These elites were instrumental in administering the colonies. For example, some social elites among the Sinhalese ethnic group in Sri Lanka benefited from tea plantation agriculture before, during and after independence. They replaced the British in governing post-independence and were ineffective in uniting the

Figure 2.2 Jantar Mantar

new nation-state. In other contexts, wealthier or privileged groups within the colony used their advantages to write about and develop new forms of anti-colonial consciousness among the population, such as the revolutionary writer and thinker Franz Fanon in Algeria. Fanon was an academic who critiqued French colonialism to show how the colonized were expected to become like the colonizer (i.e., dress, food, work, etc.), and whose writings on race and colonialism were impactful in the 1960s onward (see also Fanon 1963).

While many other ideas emerged from the Enlightenment, in what follows we focus on the concepts of biological science, merchant capitalism and the creation of republics to understand how intersectional hierarchies emerged during the period of European modernity. We trace their relevance to contemporary economic development through the ways in which Enlightenment values continue to shape paradigms of progress, reason and governance. We cover these organizing principles briefly below, but we reinforce and draw on them throughout the book.

Science: the development of modern science in the Enlightenment begins with Sir Francis Bacon. The application of a method of inquiry into natural and biological systems with reproducible results revolutionized science and medicine. It also professionalized work traditionally done by women (midwives, herbalists whose work was eventually criminalized) and created industries led by men. The professionalization and methodological interventions lent a veneer of objectivity to scientific endeavors, regardless of what was investigated. This included the application of Darwin's insights into the production of species, or what he called "races," to human populations. Races were differentiated on the basis of physical attributes and led to a false belief in objective classification of people into hierarchies – also known as **eugenics** or **social biology**. The differentiation of people based on their knowledge and civilization (using European standards) justified the practice of missionary work, settler colonialism, slavery, forced sterilization and genocide (Gilroy 1993). The idea that some races are

FOCUS: TEA AND SUGAR IN FACTORIES IN ENGLAND

The Industrial Revolution was launched in British textile mills on the backs of slaves in the colonies. The comparative advantage of pre-existing woolen mills in Manchester gave England an edge in the textile industry when cotton, produced through slavery, became available from its Egyptian, Indian and American colonies. Waged workers in the metropole were paid so poorly to weave cotton textiles that they could only afford the tea and sugar grown by slaves in other British colonies (Sri Lanka and Jamaica, respectively) as their daily food. The presence of cheap (subsidized by slave labor) carbohydrates and stimulants revolutionized working conditions in the factories and gave supervisors justification to push workers, some of them children, to work harder and faster, with fewer breaks (Mintz 1986).

inferior is widely denounced, but its legacies persist in both domestic and foreign policies as well as strategies of development.

Mercantilism: empires grew and spread rapidly based on a demand for resources, new markets and labor as industrialism developed in Europe. A new form of economic organization emerged as a result, one favoring trade and manufacturing. It was accompanied by a philosophy that commerce generates wealth and prosperity, also known as **economic liberalism**. As demand for workers grew in factories, the finances of the empire became more tied to productive work in factories than through taxation and tribute. A middle class emerged that was composed of neither imperial elites nor workers, and who formed the backbone of capitalism through the consumption of imported food from the colony and finished goods produced domestically (e.g., tea, clothing). This process established a global assembly line of extraction of natural resources from the colonies (wood, gold, gems and agricultural products such as fruits, sugar, cotton, tea, spices) and the development of a manufacturing sector focused on the production of finished goods by waged labor, leading to mass consumption by a growing middle class. The privatization of commonly held agricultural lands as well as the separation of spheres between men and women (Landes 1988) facilitated the extraction of surplus in new ways (not just through slavery or feudalism) and created a new class of waged workers, sowing the seeds of contemporary capitalist economies.

Republics: the idea of the republic as a sovereign state, in which citizens retain power through representation, emerged from Italian mercantilist city-states with extensive wealth, but little power under monarchal rule. Rousseau, writing on the republic in the 18th century, stated that people in enlightened societies should enter into a social contract with governing bodies. Citizens would sacrifice some rights in order to ensure security from lawlessness, but retain essential privileges, such as the right to property. These ideas were preceded by the enumeration of rights in various places that include legal and political protections against the power of the empire, such as freedom from arbitrary arrest or seizure of property. Within this framework, the rights-bearing citizen was framed through gendered and raced identities, only granting rights to land-owning men of European ancestry. Over time this evolved into other kinds of rights, such as the right to vote (limited again to land-owning men until recently) as the democratic state evolved. These ideas have their roots in liberal imperialism, which asserted that imperial acquisition and colonization would lead to justice, progress and freedom.

Contemporary nation-states, capitalism and development

The Treaty of Westphalia in 1648 ended the wars of religion in Europe and set the stage for the development of the modern nation-state centered on the sovereignty of a country's bounded territory and the right of noninterference. The contemporary nation-state is given power and authority through the notions of sovereignty based on liberalism – which incorporates individual rights (i.e., private property) and the idea that capitalism will bring prosperity to all. Liberal state sovereignty is differentiated from sovereign monarchs through the "social contract" which grants citizenship to members of the population in exchange for protection. It also allows the state to claim territory and subject life within the bounds of that territory (and in some cases, beyond) to the power of the state to manage and control its populations by more or less coercive means. Sovereignty, while having real effects, especially over the rights to foster life, is a contested social construction that produces "imaginaries" of subjection, boundedness and power in what Nyers (2006) calls the "social movement of statism." The creation of subjects who value their identity as autonomous individuals politically allegiant to the nation-state and finding fulfillment through consumption is the goal of modernity and capitalism. The rewards of embracing such values, however, were largely realized by social, racial and economic elites who constructed nation-states for their benefit. As demonstrated by indigenous resistance and political action for alternative sovereignties and economies by activists such as the Land Workers Movement (MST) in Brazil, liberal state sovereignty does not completely control all people and territory in order to facilitate capitalist development.

Many scholars identify the Enclosure Acts in Great Britain in the 1700s and 1800s, which privatized common lands and forced thousands of peasants off the land, as a pivotal moment in the development of modern capitalism. **Enclosure** consolidated farmland and placed land in the hands of a few and induced thousands of workers, now without the means to produce for themselves, to work for wages in factories. This spatial shift in landownership facilitated and paralleled the transition from agrarian, feudal (or otherwise "traditional") societies toward an urbanized, rationalized capitalist society structured politically through the nation-state and its scientific management of resources and population (Foucault 1978). According to David Harvey (1990, 12), "Scientific domination of nature promised freedom from scarcity, want and the arbitrariness of natural calamity." Modernist assumptions about the separation of nature and society normalized new allegiances to the state and its guarantee of food security through innovations in agricultural science, rather than through traditional agricultural labor. This practice of enclosure continues today in the form of nationalizing natural resources, public–private partnerships and land grabs.

While the process has been ongoing, and the path was laid through colonization, privatization, migra-

FOCUS: GENDER AND LAND GRABS

Land grabs are illegal or quasi-legal acquisitions of territory by corporations or governments, typically in a country or territory (space of dispossession) other than the acquirer's own (usually a zone of accumulation). Land, usually in large pieces, is acquired through direct investment, financial speculation or foreclosures on private land or with the seizure of publicly owned lands that are rich in untapped resources, such as mineral wealth or fertile soils for biofuels. In this way, land grabs work like colonialism with the same results in terms of dispossession of public lands, displacement of the rural poor and loss of control of resources for indigenous people or governments (Borras and Franco 2010). Verma (2014) argues that land grabbing, while ostensibly gender blind, has profoundly gendered effects on the population involved in both the accumulation and the dispossession. She writes that "land grabs are power relations between and among men and women, about property (68). She asserts that these power relations mirror those in the wider society, with the end result being "increased accumulation and concentration of land by powerful corporations, investors and individuals" (68), usually elite men.

tion and waged work, the **Industrial Revolution** was thought to gather steam in the 1850s in Europe, and was fueled by the colonial acquisition of labor, land and resources through colonialism. The accumulation of capital resulting from land theft and slavery made possible the massive investment required for industrialization. Science, capitalism and the newly emerging republics worked together to promote a new form of economic organization centered on manufacturing and the development of a middle class. Labor movements such as those pursuing anti-child labor laws appealed to the social contract with the state in its role of granting rights to citizens against exploitation from industry. Capitalists, who profited from the labor of vulnerable populations, resisted labor reform. They also worked to structure an industrial economy characterized by divisions of labor between home and workplace, as well as alienation from the means of production. This paved the way for women's unpaid work in the domestic sphere, which included the maintenance of both the home and the bodies of workers, to be part of the equation when extracting surplus from waged work. It also cemented a dependence on consumption (i.e., buying food instead of growing it) that led to the creation of economies based primarily on high mass consumption. All productive activities were controlled either by the corporation or by the state, thus perpetuating a process of deskilling, over-specialization and dependency.

Post-World War politics

World Wars I and II in the early 20th century saw the fall of the world's largest empires, the Ottoman and the British, respectively. The collapse of imperial order consolidated the creation of independent states composed of rights-bearing citizens. The newly developed democratic states and republics, however, granted rights unevenly, resulting in the differentiation of rights (civil, political and social) based on race, gender and class hierarchies. The modern Middle East – largely a British creation based on access to resources in the wake of World War I – set the stage for the conflicts of World War II, the Cold War and the contemporary War on Terror (Anderson 2014). World Wars I and II, among many things, created new European states, and a realignment of power between Axis and Allied powers – including the expansion of the Soviet Union in the interwar period. The United States and the USSR were allies during World War II, but then diverged based on radically different economic and political ideologies after World War II, thus fomenting the **Cold War**.

Capitalist economic development as a strategy to recover from World War II emerged and became used as a way to contain the "communist threat." This effort was led by the United States through the Bretton Woods agreement and was meant to rebuild Europe. The US Marshall Plan provided extensive loans to many European nations after World War II to improve industry, remove trade

FOCUS: COMMUNISM IN CONTEXT

Communism developed from the writings of Karl Marx and Friedrich Engels, who critiqued both feudalism and capitalism as systems of class oppression. The Bolsheviks in Russia and followers of Mao in China took different approaches in an effort to challenge massive inequalities between wealthy elites and the poor. While the tenets of communism in these countries discursively focused on the people and workers as equal members of state-based communalism, the implementation and experiences of these beliefs were much different. State-based reforms in China and the Soviet Union developed into systems of state control that limited freedom of the press and individual mobility outside of communist spaces. While these countries provided benefits to the people through healthcare and full employment, many workers experienced abuse or ill health as a result of their work environments. Capitalist states saw communism as a threat to economic exchange and accumulation in the private sector.

barriers, and prevent the spread of communism. The US sought to incorporate newly independent states – formerly the "colonies" – into capitalism through loans and development assistance. Countries and their governments were at the center of both provisioning and receiving development aid and assistance. Nonaligned states (also known as the Third World) were targeted for inclusion in the capitalist economy through economic development. Many newly independent states were incorporated into the capitalist economy when they turned to the wealthier countries (e.g., the United States) for loans, thus committing themselves to capitalism through the mechanism of debt.

Notable exceptions to this strategy included Cuba, which isolated itself politically and economically, while receiving aid from the Soviet Union and other countries. The debt trap left many struggling young countries desperate for cash, and willing or coerced into trading on their narrow resource base, such as petroleum or sugar, often replicating the colonial arrangements from which they had sought to be free. Debts take a multitude of forms, some of which arrived with emergency loans to cover basic expenses, which were then accompanied by conditionalities to restrict funding on other priorities, such as education. Other forms include **colonial debt**, which is the payments that former colonies make in various forms to their former colonizers for infrastructure built during occupation.

> After the successful revolution of Haiti to overthrow slavery and French colonialism, Haiti was forced to pay France 90 million gold francs in reparations. This was not just a perverse reversal of obvious justice, where the enslaving colonial power forces compensation for the loss of illegitimately held land and human beings from the victims of its slavery, it was also a devastating economic imposition that some experts hold "played a large part in the Caribbean country's subsequent descent into stark poverty and under-development."
>
> (Sargent 2012, 200)

The Bretton Woods agreement in 1944 articulated the economic vision for the postwar era, defined through inter-state economic cooperation. The flow of capital and goods was to be regulated at the global level under the newly created institutions of the International Monetary Fund (IMF), the World Bank (formerly the International Bank for Reconstruction and Development) and the International Trade Organization (ITO), which later became the World Trade Organization (WTO). The World Bank was largely established for postwar reconstruction, with the needs of the former colonies registering only as an afterthought. Today the World Bank is one of the most powerful development institutions and its raison d'être is to finance capitalist economic development projects, largely in former colonies and post-Soviet states. The IMF was created to establish stability in exchange rates in currencies between countries, and to stabilize the global economy through making loans to member countries in times of economic crisis due to debt.

In 1945 the United Nations (UN) (formerly the League of Nations) was established by the Allied forces to promote international cooperation in the era of postwar peace. Its primary functions are to ensure security, foster economic development, define and defend human rights and provide humanitarian disaster assistance. The UN is tasked with assessing, interpreting and communicating the social and economic needs of its member countries. For example, one of its primary agencies is the World Health Organization, which is responsible for monitoring, reporting and resolving threats to human health, such as the spread of the Zika virus. The World Bank, the UN Development Fund, UNICEF (United Nations International Children's Emergency Fund), UN Women and the International Development Association (IDA) are administered by the UN under the umbrella of the World Bank Group. The UN seeks to end extreme poverty and to promote prosperity through providing loans and assistance to developing countries, former colonies and newly independent states. The primary target is agricultural and rural development, infrastructure (e.g., roads), industrial projects and governance. The UN is headquartered in New York (with additional main offices in Geneva, Nairobi and Vienna) and the World Bank is located in Washington, DC. The location of these headquarters exemplifies the disproportionate influence of powerful countries within the UN and other supranational organizations.

Much of the world's economic policy is thus informed by Western interests as well as defined by capitalist economic philosophy. In terms of economic policy, there are two prevailing trends. Polanyi (1944) referred to these as a double-movement, arguing that capitalism needs regulation to temper its tendency toward crisis. Deregulation often begets situations of regulatory reform, while regulation foments resistance to restriction, hence the "double" movement. This complex interplay between market freedom and policy formation ensures the existence of the liberal state, because, according to Polyani (1944), neither markets nor states can exist or persist without each other. Between the 1940s and the 1980s, the United States (and much of the capitalist world) operated under a **Keynesian** regulatory environment, in which state governance involved itself in economic policy. Keynes' economic philosophy was to use the state to protect citizens from the negative externalities of capitalism, as occurred in the Great Depression of the 1930s (Mann 2013).

During this time, the Cold War shaped the geopolitical environment, in which the United States and USSR fought proxy wars as a way to implement their own forms of economic development. The United States was central to the instigation of wars in Latin America to stop the rise of socialist democracies in the western hemisphere. In the 1980s, a neoliberal philosophy emerged, resulting in widespread deregulation and changes to currency and corporate policy. This came to be known as the Washington Consensus, which assumed that free market philosophy would ensure prosperity and public goods, rather than state involvement. The Washington Consensus imposed its economic policies through mechanisms put in place by the IMF and World Bank, such as the implementation of austerity through externally imposed constraints on the economy, known as structural adjustment. This economic philosophy and practice opened the doors to globalization and the spread and expansion of liberal markets to the rest of the world. The connection between economic development and globalization continued into the 1990s and expanded further after the fall of the Soviet Union in 1991.

In the conventional wisdom of economic development, a country proceeds from very basic economic activities, such as agriculture and natural resource extraction, to a knowledge economy based on technology. This general path proceeds from modernizing investments in agriculture and natural resources (also referred to as a primary sector) that free farmers and peasants to migrate to work in an industry (secondary sector), usually manufacturing. Many countries start out with a relatively large unskilled labor force producing inexpensive consumer goods. As the industrial sector becomes more specialized (e.g., automobile manufacturing), laborers become more skilled and the country gains more wealth, a service economy (tertiary sector) emerges to meet their needs for prepared foods, healthcare, education and other essentials, as well as luxuries. In this model of development, a large middle class is intended to emerge as a result of this prosperity. As the service sector expands and wages increase as a result of worker organization and the increase in the cost of living, the industrial sector is exported to other newly independent countries or transitional economies, such as India or China. As the service sector comes to dominate the economy in the late 20th century, a new sector emerges through the knowledge or creative economy. Work in this sector is characterized by science, technology, engineering and programming. This model is largely based on the experience of Western economies in the 20th century, and currently dominates much of economic development thought. This segmentation of the economy, based on previous inequalities and the lack of a level playing field that is supported by supranational organizations, paved the way for the global assembly line, which is differentiated on race, class, gender and national origin.

Gender as a development concern was not on the radar of the UN and other associated development agencies until 1975. The 1970s have been named as the decade of women, during which the UN began observing and responding to issues specific to women in the development process. The WID, WAD, GAD, WED and DAWN formulations (see Chapter 1) emerged over the next decade as a way to understand how to address women's needs in a gendered development context. By the end of the 20th century, the UN identified a set of Millennium Development Goals to address extreme hunger and poverty and to specifically attend to women's primary needs in terms of equity, empowerment, maternal health, child mortality and HIV/AIDS. Gender and health have become integral to development efforts, including humanitarian relief, construction of sanitation infrastructure and improving access to medical care. Public–private partnerships, using capital from corporate profits, often channeled through nongovernmental organizations such as the Bill & Melinda Gates Foundation, have grown along with the expansion of government-based development organizations. **Neoliberalism** is the form of liberal economic policy which uses state policy measures to ensure the accumulation of capital in the private sector (Harvey 2007). The emphasis on the use of private capital through public policy to create prosperity (as opposed to earlier models focused on state-led development strategies) normalizes new paradigms of social entrepreneurialism, military intervention, free market capitalism, microcredit and land titling. Žižek (2006) argues that capital must be stolen through colonialism before it can be given back through aid or humanitarianism. Likewise, the new private capital norm in development provides aid but does not question from where and why poverty appeared, and perpetuates the theft of capital and capabilities from the poor.

The ideological framework behind including women/gender as part of international capitalist development was to advance platforms for empowering the world's least empowered people, such as women. This approach, however, did not ask important questions about how and why women were in positions of precarity and vulnerability in the first place. International aid organizations often imposed their own normative visions of feminism and empowerment on women and other recipients of aid. Therefore, these organizations did not seek to understand women's own methods of gaining power and influence in their communities. These efforts often backfired by alienating women from traditional sources of support, leaving them worse off than they were prior to development intervention. Moreover, these approaches sought to impose **liberal** logics on people who did not view individualism or private property as a core value, nor derived satisfaction from profiting at the expense of others. The belief that humanitarianism and development could bring freedom and prosperity masked the efforts to integrate the most vulnerable people into markets and make them efficient workers. Belief that the free market provides collective goods also fails to attend to precarious places with fragile resource bases, where individual gain can lead to social and economic disasters.

Feminism, intersectionality and capitalism

Feminism has had a significant impact on social life since women's movements in various countries challenged the patriarchal status quo in the 1970s (Bondi 1990). Suffrage movements, inspired by liberal feminism, began globally in the 19th century, but did not see results until the mid-20th century in most places (Landes 1988). Liberal feminism sought to increase the number of women in economic and political structures as well as pushing for gender pay equity. Liberal feminism is fundamentally based in rights, and demanding equal access to rights afforded to men (Scott 1996). In other words, equality meant that women existed as a singular category and anyone in that category should be able to receive the same social benefits as men. This approach ignores the way in which equality with men erases important gender differences that may be relevant to women actually achieving those rights. For example, childbearing and childcare responsibilities may impact women's employment, thus impacting wage equity. This approach also elides differences between groups of women who are not of elite classes, such as women of color, poor women and women in postcolonial contexts, and who may be oppressed differently by patriarchy and dominant constructions of masculinity (Mohanty 2005). In practice, liberal feminism (now often referred to as white feminism) advanced the interests of white, usually middle- and upper-class, women at the expense of queer and women of color and some intersectionally gendered men. This also had the effect of sparking lively debates among feminist scholars about what constituted the category of "woman" and which women could receive equal rights and protections under liberal feminism (Riley 1988).

Radical feminism emerged in response to the failure of liberal feminism to incorporate the needs and realities of women of color and postcolonial contexts. Radical feminism advanced the idea that resistance to **patriarchy** (or male supremacy) in all its forms, including capitalism, the church or colonialism requires restructuring society in ways that met the needs of women-identified people for social, economic and political security (McDowell 1993). The political process through which this agenda is advanced is not through the extension of rights, but through changing institutions to respect alternative forms of gender relations and identities, raising awareness of and condemning sexual assault, rape and domestic violence and recognizing multiple forms of womanhood or femininity. Within the same time frame Marxists and materialist feminists contributed to understanding how categories of identity are produced through the ideological apparatus of capitalism, which requires an underclass to produce profit (MacKinnon 1982). The underclass is composed of people who are paid less than others in similar workplaces and professions, and feminist contributions to Marxism underscored how this often includes men and women of color, indigenous people and/or migrants who are oppressed in multiple, intersecting ways (Anthias and Yuval-Davis 1983). Mohanty (2005) asserts that feminism must incorporate an explicitly anti-capitalist orientation that denaturalizes the social relationships that shape contemporary life, including anti-immigrant sentiments, nationalism, militarism and corporate culture.

Direct action, boycotts, protests and subversive art characterize many of the methods used by radical feminists to incite a social revolution around gender and other forms of intersectional identity (Combahee River Collective 1977). These political interventions created philosophical and legal spaces for the concept of **intersectionality** to emerge and take on a political life (Crenshaw 1991). Intersectionality emerges in black-feminist scholarship and refers to overlapping and interlinked forms of oppression based on multiple sociopolitical categories such as race, socioeconomic class and sexuality. For example, intersectionality was important for legal cases where a person experienced multiple forms of discrimination (e.g., being barred from certain work because of one's gender *and* based on one's skin color) (Cho, Crenshaw and McCall 2013). Intersectionality carries forward the idea that women do not constitute a coherent group across space and time, and are not oppressed by systems of power in exactly the same way (Hill-Collins 2000).

While intersectionality is a conceptually useful tool for understanding difference and inequality in its myriad forms, it is not without critique. The first and most pressing critique relates to the uneven treatment given to different axes of identity within feminist scholarship. While race and gender receive a great deal of attention, less attention is paid to issues of sexuality or class (Yuval-Davis 2006). Following from this, Yuval-Davis asserts that within the rather structural framework of the metaphor of intersections, deciding upon the optimal number of intersections to be considered at any given time is impossible; nor is it possible to assess an infinite number simultaneously or how to prioritize any one to consider. Finally and relatedly, it is often presumed that black women are "quintessentially intersectional, while others are not" (Brown 2012, 545). It is also often the case that people come by intersectional identities as if by magic, without consideration of how those particular identities are produced in the first place. Seeking a way forward methodologically with this flawed but still useful heuristic is essential for the work of understanding the experiences of differently situated men and women who occupy and reproduce varying forms of identity.

Postcolonial feminism developed in response to the overt focus on white and Western women in earlier feminist interventions, which overrepresents and often misrepresents the experience of women living outside of zones of accumulation. Postcolonial feminists aim to call attention to the ways in which women are differentiated from each other, and thus require different forms of intervention in the achievement of basic rights (Mohanty 2005). For example, reproductive rights might mean access to contraception or abortion for white women, but for black or indigenous women, being allowed to give birth and raise children in healthy environments may be a far more basic need – and in some places, this need is not currently being met (Ross et al. 2016). Postcolonial feminists call attention to the fact that under/misrepresented women do not have full access to the category of "human" through the ongoing legacies of racism, colonialism and genocide (McKittrick 2014). Gilmore (2002, 16) writes that racial–sexual categories produced through imperialism create differences (white middle class, black working class, indigenous poor, etc.), which are then displaced into hierarchies "that organize relations within and between the planet's sovereign territories." This hierarchy thus structures the global economy along racial–sexual lines and creates "naturalized" justifications for inequality that manifests in the unpaid labor of women, gendered wage gaps and the disposability of workers (Wright 2006).

Engendering development

We assert three main points that will be carried forward in this volume. The first is that the imperialism of the modern era created categories of difference between humans that had not existed before. These are specifically modern constructions that powerfully shape social, economic and political realities. They are not the result of "natural" or biological differences, but rather individuals working within institutions with the power to subordinate others, systematically creating and enforcing these divisions socially, politically and legally. Secondly, the differences made visible by imperialism remain evident today because the exploitative and extractive political project of colonialism is still alive and well. To seize the wealth of colonized people, they first had to be created as inferior to the colonizer, who was almost always a white European, upper-class elite male. This process is not unique to the British Empire, but was similarly orchestrated by other empires, such as the Japanese who saw themselves as racially superior to other Southeast Asian people. The categories that emerged in relation to European identity shaped the development of capitalism and continue to be enrolled in the production and reproduction of the global economy today. Third, we assert that development is an ongoing process of enclosing and extracting resources, especially human labor. The categories of identity that imperialism produced continue to be used to structure the capitalist economy in order to produce surplus. As we will discuss in the chapters that follow, development works through the apparatus of capital investment, population management, humanitarianism, technology transfer and conflict.

In summary, this chapter examined development through the lens of three central themes. First, **inequality** is essential for capitalist economic development to work. Inequality is created through the displacement of difference into hierarchies and through the creation of "others," who over time and space and in various ways are not seen as fully human. Second, when people (immigrants, women, nonwhite men, indigenous, the poor) are not seen as fully human, they are not given full access to rights, and they are exploited with low or no wages for their labor. Third, we take an intersectional approach to understanding the creation of these categories of identity that are used to oppress and exploit. The differences that are created along the lines of gender, race, class, ethnicity, etc. between groups of people, particularly in the context of scarce resources, drives the accumulation of surplus in the hands of the already wealthy. This generates an additional set of inequalities and vulnerabilities to exploitation and the withholding of rights in the state system, especially if they are forced to migrate to places where they may lose citizenship rights. This is a cyclical process upon which capitalism and development thrives. We now turn to analyzing the links between development, capitalism and humanitarianism in the production and maintenance of international business.

Recommended reading

Poverty capital, Ananya Roy; *Confessions of an economic hitman*, John Perkins; *Sweetness and power*, Sydney Mintz; *The value of nothing*, Raj Patel; *The conquest of bread*, Petr Kropotkin

Recommended viewing

Rabbit-proof fence; *Loving*; *Mardi Gras: made in China*; *This magnificent African cake*; *Royal affair*; *Earth*; *Food chains*

Questions for discussion

What does imperialism look like today? How is your identity shaped by empire (think relationally)? Why is empire often venerated, rather than vilified?

References

Anderson, S. (2014). *Lawrence in Arabia: war, deceit, imperial folly and the making of the modern Middle East*. London: Atlantic Books.

Anthias, F., & Yuval-Davis, N. (1983). Contextualizing feminism-gender, ethnic and class divisions. *Feminist Review*, 15(1), 62–75.

Bales, K. (2012). *Disposable people: new slavery in the global economy*. Berkeley, CA: University of California Press.

Bondi, L. (1990). Feminism, postmodernism, and geography: space for women? *Antipode*, 22(2), 156–67.

Borras, S., & Franco, J. (2010). Towards a broader view of the politics of global land grab: rethinking land issues, reframing resistance. *Initiatives in Critical Agrarian Studies Working Paper Series*, 1, 1–39.

Bouie, J. (2018). How the Enlightenment created modern race thinking, and why we should confront it. https://slate.com/news-and-politics/2018/06/taking-the-enlightenment-seriously-requires-talking-about-race.html. Accessed 10/8/2018.

Brown, M. (2012). Gender and sexuality I: intersectional anxieties. *Progress in Human Geography*, 36(4), 541–50.

Combahee River Collective (1977). *A black feminist statement* (pp. 210–18).

Cho, S., Crenshaw, K.W., & McCall, L. (2013). Toward a field of intersectionality studies: theory, applications, and praxis. *Signs: Journal of Women in Culture and Society*, 38(4), 785–810.

Crenshaw, K. (1991). Mapping the margins: intersectionality, identity politics, and violence against women of color. *Stanford Law Review*, 43(6), 1241–99.

Domosh, M., & Seager, J. (2001). *Putting women in place: feminist geographers make sense of the world*. New York, NY: Guilford Press.

Faludi, S. (2007). America's guardian myths. *New York Times*, September 7.

Fanon, F. (1963). *The wretched of the earth*. New York, NY: Grove Press.

Foucault, M. (1978). "Part five: right of death and power over life." In *The history of sexuality, volume one*, New York, NY: Vintage Books (pp. 135–59).

Gilmore, R.W. (2002). Fatal couplings of power and difference: notes on racism and geography. *The Professional Geographer*, 54(1), 15–24.

Gilroy, P. (1993). *The black Atlantic: modernity and double consciousness*. Cambridge, MA: Harvard University Press.

Goldner, V. (1991). Toward a critical relational theory of gender. *Psychoanalytic Dialogues*, 1(3), 249–72.

Harvey, D. (1990). *The condition of postmodernity: an enquiry into the conditions of cultural change*. Oxford: Blackwell.

Harvey, D. (2007). *A brief history of neoliberalism*. Oxford: Oxford University Press.

Hill-Collins, P. (2000). *Black feminist thought: knowledge, consciousness, and the politics of empowerment*. London: Routledge.

Landes, J.B. (1988). *Women and the public sphere in the age of the French Revolution*. Ithaca, NY: Cornell University Press.

Lewis, R. (2013). *Gendering orientalism: race, femininity and representation*. London: Routledge.

MacKinnon, C.A. (1982). Feminism, Marxism, method, and the state: an agenda for theory. *Signs: Journal of Women in Culture and Society*, 7(3), 515–44.

Mann, G. (2013). *Disassembly required: a field guide to actually existing capitalism*. Edinburgh: AK Press.

McDowell, L. (1993). Space, place and gender relations: Part I. Feminist empiricism and the geography of social relations. *Progress in Human Geography*, 17(2), 157–79.

McKittrick, K. (2006). *Demonic grounds: black women and the cartographies of struggle*. Minneapolis, MN: University of Minnesota Press.

McKittrick, K. (Ed.) (2014). *Sylvia Wynter: on being human as praxis*. Durham, NC: Duke University Press.

Mintz, S.W. (1986). *Sweetness and power: the place of sugar in modern history*. London: Penguin.

Mohanty, C.T. (2005). *Feminism without borders: decolonizing theory, practicing solidarity*. Chapel Hill, NC: Duke University Press.

Nyers, P. (2006). *Rethinking refugees: beyond states of emergency*. New York, NY: Routledge.

Polanyi, K. (1944). *The great transformation: the political and economic origins of our time*. Boston, MA: Beacon Press.

Riley, D. (1988) *"Am I that name?": feminism and the category of "women" in history*. Minneapolis, MN: University of Minnesota Press.

Ross, L., Gutiérrez, E., Gerber, M., & Silliman, J. (2016). *Undivided rights: women of color organizing for reproductive justice*. Chicago, IL: Haymarket Books.

Sargent, P.K. (2012). Debt cancellation as reparation: an analysis of four cases. *Armenian Review*, 53(1–4), 193–205.

Scott, J.W. (1996). *Only paradoxes to offer*. Cambridge, MA: Harvard University Press.

Smith, S. (2012). Intimate geopolitics: religion, marriage, and reproductive bodies in Leh, Ladakh. *Annals of the Association of American Geographers*, 102(6), 1511–28.

Rizzo, T., & Gerontakis, S. (2016). *Intimate empires: body, race, and gender in the modern world*. Oxford: Oxford University Press.

Verma, R. (2014). Land grabs, power, and gender in East and Southern Africa: so, what's new? *Feminist Economics*, 20(1), 52–75.

Wright, M.W. (2006). *Disposable women and other myths of global capitalism*. New York, NY: Routledge.

Yuval-Davis, N. (2006). Intersectionality and feminist politics. *European Journal of Women's Studies*, 13(3), 193–209.

Žižek, S. (2006). *The parallax view*. Cambridge, MA: MIT Press.

3 The business of international development

Introduction

Examining the business of development elucidates the ways in which international humanitarian and development assistance operates and, in many ways, contributes to the generation of wealth for an increasingly small group of already wealthy people on the backs of the impoverished many. In this chapter we examine the differences and similarities between humanitarian assistance and economic development. We follow this with a critical analysis of the profession of international development. The business of international development, similar to other businesses, relies on advertising and fundraising. Celebrity advocacy and advertising have become increasingly integral for promoting the concept of humanitarian assistance, economic development, representing what "need" looks like, and helping organizations to generate funds. The remaining sections provide a critical overview of online fundraising campaigns that mimic popular online shopping platforms. Additionally, economic intervention programs intended to assist poor and exploited laborers are used as examples of both the promise and pitfalls of various attempts to work within and challenge existing capitalist development/business frameworks.

Humanitarian assistance and economic development

Both humanitarian aid and economic development organizations situate their work as a form of assistance to recipient countries and communities. Each organization endorses its own ideas for how to implement economic or resource aid (Schaaf 2013). While these organizations are often perceived as "assisting others," the structures of humanitarian assistance and economic development increasingly operate within neoliberal capitalist logics. Consequently, international aid and development has transitioned into a multi-billion-dollar industry that includes both for-profit and not-for-profit organizations. Some donor organizations, countries and individuals profit or generate significant income through the processes of implementing economic development programs. Geopolitically, most zones of accumulation leverage international assistance and development projects to garner support, gain allies and influence societies to be "more like" a zone of accumulation. Development often mimics colonial processes that once forced places to reflect the social, economic and political beliefs of colonists, and now expects countries to conform to (and become dependent upon) global capitalism. For example, assistance and development programs have

FOCUS: HUMANITARIAN AND DEVELOPMENT BUSINESSES

Some development organizations are supranational and provide significant sources of funding through grants and loans to both government and nongovernmental development or humanitarian aid organizations, such as the United Nations (UN), the World Bank, the International Monetary Fund (IMF), the Global Fund and Asia Foundation, along with charitable and philanthropic organizations such as the Bill & Melinda Gates Foundation, the Open Society Foundation and the Aga Khan Foundation. International nongovernmental organizations (NGOs) are another category of development institutions which include secular and faith-based organizations such as Oxfam, CARE International, Save the Children, Heifer International, World Jewish Relief, World Vision, Catholic Charities, International Islamic Relief Organization and Christian Aid. Many governments have dedicated aid/development organizations, such as the United States Agency for International Development (USAID), Japan International Cooperation Agency (JICA), China Development Bank, UK Department for International Development (DFID), Australian Aid, Canadian International Development Agency (CIDA), Swedish International Development Agency (SIDA), and state consortiums such as the European Commission, and OPEC Fund for International Development (OFID).

been used as a mechanism for exporting genetically modified crops that render food systems dependent on seed purchasing rather than being able to reproduce crops without purchasing seeds, such as seed saving. In other examples, assistance and economic development have paved the way for the creation of low-wage factory employment in order for corporations to transfer jobs from countries with higher wages to countries with lower wages in order to maximize or increase profits for shareholders.

Brockington (2014, xxii) identifies development as what an individual or community does and can do by themselves, while "humanitarianism requires a needy *other* . . . The history of humanitarianism begins with the recognition of the humanity of distant strangers." In other words, assistance is premised on economic inequalities and the "needy other" who is incorrectly represented as without agency and therefore "not qualified" to speak on his/her own behalf. Despite this fundamental difference, humanitarian and development work merge. For example, as discussed in Chapter 1, humanitarianism remains a common precursor to economic development in the aftermath of dispossession, conflict or independence from a colonial power. Humanitarian organizations arrive in various locations to attend to basic human needs (food, water, shelter), which have been disrupted due to conflict, environmental disaster or economic devastation (see Chapter 9). Economic development projects follow humanitarian assistance and many humanitarian assistance organizations manage economic development projects. Similarly, economic development organizations, whether they are state-based, nongovernmental or supranational, often merge **humanitarian assistance** with economic development programs; humanitarian assistance is generally expected to occur within relatively short time frames in order to provide for acute needs and basic human survival, while **economic development** operates on longer time frames toward sustained economic growth and stability.

International assistance is dominated by development organizations representing various countries and many NGOs, most of which are not-for-profit. There are also a growing number of for-profit companies that capitalize on development funding allocations from countries and large donor organizations. For example, international security and military corporations, construction companies and logistics companies, as well as for-profit development organizations, receive funding from nonprofit organizations to assist with the implementation of projects in different countries. In the United States, "beltway bandits" is a common pejorative term used to describe for-profit organizations that set up offices near the "beltway" (highway) near Washington, DC in order to be strategically located close to USAID, the US Department of State and the US Department of Defense (DOD), to secure government contracts for international aid/development work. The relationships between government funds, nonprofit and for-profit businesses operate as a web of different organizations and institutions, which turn public money (i.e., taxes) into "private accumulation (including on the part of so-called nonprofits) in the name of development" (Roberts 2014, 1046). In this way, citizens' taxes are allocated to international assistance and development while simultaneously many companies are also profiting from these efforts, whether or not the projects are successfully implemented or result in positive change in a given country (Fluri and Lehr 2017).

International aid and development workers

The business of economic assistance and development includes the professionalization of a skilled humanitarian aid and development workforce, and often does more for the development worker than it does for those receiving the aid. Many individuals are drawn to work in the world of aid and development for a variety of reasons. Malkki's (2015) research on Finnish assistance workers found that the "need to help" occurred in tandem with the desire for international travel and to live in places that were much less ordered, disciplined and restrictive than their home communities. Malkki identifies aid and development workers as more focused on self-escape than self-sacrifice. "Self-escape" was identified by many as a primary motivator for many international aid/development workers to leave home and work abroad. The "personal desires" of these workers to break away from their daily routines and predictability of safe living and working environments "were definitely factors in their decisions as to whether to accept an international mission or not" (Malkki 2015, 11).

International workers from Europe and North American living/working in Afghanistan revealed similar motivations for working in this location, along with identifying that they enjoyed "the thrill" and excitement of living in a conflict zone (Fluri 2009). Several internationals working in Kabul, Afghanistan (2006–12) who previously worked in other war-impacted locations (such as Bosnia, Rwanda or Iraq) also described being "addicted" to conflict zones and feeling more at home in these places than in their home countries. Their "addiction" was associated with the sense of adventure they experienced living in a space of heightened security along with the lack of oversight or law enforcement governing their individual behavior (Fluri and Lehr 2017, 36). In other cases, workers described spaces catering only to international workers (such as bars and restaurants) as similar to college fraternity parties, or an experience of freedom from the rules and norms of their home societies. In many respects, international aid and development workers remain a cosmopolitan, elite and highly educated group of privileged individuals, some working towards improving societies, others seeking economic or professional gain. The worst international workers, particularly in conflict zones or in the aftermath of environmental disaster, use their position of privilege and power to extort or abuse the most vulnerable of the populations they are employed to assist and serve (see Chapter 9).

Aid/development has historically been and continues to be male-dominated, though the number of female international workers is increasing. Additionally, international workers are expected to have a certain level of education and expertise; therefore, aid/development work has become its own profession and many institutions of higher education offer professional master's degrees in international development. Professional international aid/development workers are generally from economically privileged countries with better access to higher education and common interests in international assistance work. Development experts further institutionalize the practices of development and exemplify the unequal relationships between donors and recipients (Kothari 2005, 426). Therefore, the development professional is identified as an "expert" not necessarily because of their level of education, knowledge or effectiveness, but mainly based on their home location which legitimizes their authority and expertise. Subsequently, it reinforces the binary classifications used to quantify and qualify developed and underdeveloped spaces (Kothari 2005, Bhabha 1994). The professionalization of development includes new positions that seek to improve women's lives and mitigate gender inequality; "gender advisors," "gender specialists" or "gender experts" are some of the job titles associated with these positions.

Gender advisors or "experts" often prioritize increasing women's economic participation in the paid workforce rather than tackling the broader and more difficult issues of unequal power relations associated with intersectional gender roles/relations, ethnic or racial divisions, or systemic economic inequalities (Kabeer 1994, 2005, Kothari 2005). Furthermore, many gender "experts" work in countries where they do not know or understand local sociocultural norms that shape gender roles and relations. The overwhelming majority of social scientists who study gender continue to underscore the importance of understanding how social and cultural mores, economics and politics both shape and are shaped by gender roles and relations. Additionally, Judith Butler's (1990) groundbreaking work on gender performativity shows that it is the *doing of gender* that is the mechanism of its construction and reproduction through repetition. In other words, gender roles (like other social mores) are learned rather than biologically conditioned behaviors. Gender occurs by *doing*, and this *doing* is mediated by spatial, social, cultural and political contexts of daily life repeated over time (Nelson 1999). Therefore, without an understanding of how gender **performativity** (or the *doing* of gender) operates in a given place, gender advising, or expertise, is often based on donors' rather than the recipients' own understanding and conceptualizations of gender roles/relations.

Concepts of gender, gender mainstreaming and assistance remain dominated by Western liberal ideologies scaled to represent "universal" rights, rather than addressing gender roles and relations of the sociocultural context within which they exist. For example, analyses of Iranian Islamic feminists identified that these women were criticized by US-based feminists for failing to take up certain issues (such as homosexuality and women's autonomy), while mainstream liberal feminists in the US have not demanded major alterations to existing political or economic systems.

> What liberal feminists have not called for is a change in the system of taxation and in development policy that would alter American foreign policy and the distribution of wealth, transforming the lives of low-income women in the United States and elsewhere.
>
> (Moghadam 2002, 1159)

Moghadam's argument identifies that Western liberal feminists' successes fit within the socioeconomic system of capitalism (Fraser 2013). Conversely, capitalist forms of Western liberal feminism do not always work when implemented in other socioeconomic or cultural contexts. Thus, when international aid/development workers attempt to socially engineer gender roles, they are often met with resistance, because they are seeking to change gender roles and relations to fit within a larger framework of global capitalism, rather than work within the existing place-based socioeconomic structures to improve the lives of women (and men).

In aid/development spaces organizations from Europe or North America privilege international workers with English language proficiency and college and graduate degrees. These skills allow aid/development organizations to craft their newsletters, brochures and proposals with high standards of English, which remains the dominant *lingua franca* of international development from these geographic areas. The ability to communicate (both verbally and in written documents) in English remains significant for finding work and upward mobility within the hierarchical structures of large international aid and development organizations headquartered in North America and Europe. Additionally, the professionalization of development work has included relatively high salaries for international aid/development workers. Working for a large and well-funded international aid or development organization can yield a lucrative income and other economic benefits. Many professional development workers, especially those who work "in the field" or "on the ground" in a foreign country earn significantly higher salaries than their coworkers who are citizens of the host country. In a multitude of cases, local staff salaries are fractional compared to those of their international coworkers. In spaces of conflict or political unrest, some international organizations offer their international staff "danger or hazard pay," which includes additional salary to offset working in an insecure environment (Fluri 2009). Danger/hazard pay rates change based on the perceived level of threat, and therefore vary between approximately 20 and 50 percent of one's base salary. For example, in Afghanistan in 2006, some organizations' international workers were making $250/hour and local Afghan workers in the same organization were making $250/month.

In several cases, young professionals trying to improve their career choices or build their résumés will work in locations identified as difficult, insecure or in the aftermath of conflict/disaster. Working in a location considered "undesirable" or "difficult" is one method for improving one's job prospects for assignments in more "desirable locations" (Fluri and Lehr 2017). In addition to professionalizing much of the global aid and development workforce, international organizations rely on various public and media forums to "sell" the idea of assistance and fundraise donations for their projects and programs. Similar to other businesses, humanitarian assistance and economic development advertising includes celebrity advocacy and sponsorship. The following section provides an overview of consumer-based advertising, fundraising and celebrity advocacy.

Celebrities and humanitarian and development assistance

Celebrity advertising for different economic development and assistance organizations is becoming more common. For example, the #AltGift commercial for Heifer International presents various celebrities explaining how to use Heifer's gift-giving program to assist families in developing countries. Despite the high profile of celebrities and their visibility as charitable donors, they rarely discuss or address the root causes and driving forces behind gender, sexuality, class and racial inequalities globally and within various geographic locales.

Charitable giving by celebrities and philanthropists identifies their acts as generous or altruistic without addressing or identifying how these individuals and organizations are embroiled in the business and branding of humanitarian and development assistance. Celebrity giving remains a significant and integral method of generating funds for charities along with international aid and development organizations. Most mainstream popular media outlets represent international humanitarian or development organizations as altruistic and providing assistance to economically disadvantaged individuals throughout the globe with the money contributed by wealthy individuals and groups. For example, Collier (2007) identifies the ways in which privatization and neoliberal economic structures have helped to turn charity work and economic development into a thriving business.

Therefore, we ask the question, if celebrities and other wealthy philanthropists, in addition to the extensive number of governmental and nongovernmental organizations, give and generate funds towards poverty alleviation, why do poverty and economic inequality persist? While charitable giving and other forms of development aid remain an important part of assistance, the gap between wealthy and poor countries and individuals continues to widen further every year. For example, the world's eight wealthiest (mostly white) men have amassed more wealth than half of the world's population combined (Elliott 2017). Thus, the framing of charity and need relies on existing gendered tropes that situate women of color as oppressed, poor and in need of rescue by a white, wealthy, female or male guarantor. In this section we remove the pretty veneer of charitable giving to illustrate how aid, development and charity reaffirm rather than challenge economic inequality.

Much of international humanitarian and development assistance uses consumer capitalism as a framework for fundraising campaigns. (PRODUCT) RED exemplifies a project designed to link consumer spending with assistance, i.e., combating HIV/AIDS in Africa. Designer and well-known companies such as Gap, Apple and Armani branded their products with the RED label. They advertised that a percentage of the profits on each (PRODUCT) RED labeled item would be sent to fight HIV/AIDS in Africa, thus encouraging consumers to purchase these items. Bono (lead singer of U2 and self-identified humanitarian/development celebrity) endorsed this project, along with international development "experts" turned celebrities Jeffrey Sacks and Paul Farmer (Richey and Ponte 2011). By many measures (PRODUCT) RED was a successful program; however, it distanced corporations from being part of the problem of global economic inequalities by situating them as a solution. For example, this program focused on profit sharing for charity, but did not discuss or address the treatment of the laborers making the RED labeled products.

(PRODUCT) RED, along with similar programs, incorrectly elevates consumption to an act of morality on the part of the consumer. This and other programs solicit individuals to "help" by way of consuming products. Consumer-based giving feeds into saving and hero myths that perpetuate the idea that economic development/assistance is the best or only path toward helping others. Purchase-based programs like the (PRODUCT) RED campaign ask consumers to "buy that celebrity-endorsed T-shirt, but do not ask too many questions regarding the sweatshops" that produced that shirt (Hasian 2016, 17). These campaigns focus on basic assistance while they do not address or attempt to mitigate global inequalities endemic to consumer-based and free market-driven capitalism. In many respects, these campaigns use visual advertising techniques that reinforce dichotomous differences between the wealthy (those giving/purchasing) and poverty (those receiving assistance). Wealth is continually marketed as aspirational rather than exemplifying economic inequality.

Organizations such as Mercy Corps, Oxfam, Save the Children, CARE International and Heifer International have incorporated online shopping platforms as one of their fundraising techniques. These donation techniques simulate other online consumer-based shopping environments, with the use of bright colors, images, videos and text. Women and children of color comprise the overwhelming majority of individuals pictured on these sites, thus representing a gendered and racialized vision of need. Men are much less commonly shown as compared with women and children. When men are pictured they are often elderly, shown in the distance rather than close-up, or are disabled. White men are not pictured, except for images of white males physically disabled in wheelchairs (Fluri 2017). The absence of white, able-bodied males illustrates how economic privilege and disadvantage are represented through different racial and gendered bodies (Rothenberg 2008). While individuals of various skin colors, ethnicities, genders and locations experience wealth, economic opportunity or poverty, these websites reinforce stereotypes that associate poverty with the black and brown bodies of women and children, and economic privilege with white skin and the male gender.

White females are often depicted as agents of care who purchase consumer-based forms of assistance. Carefully chosen images help to situate a global imagination of what need "looks like" in contrast to the bodies of (mostly) white middle- and upper middle-class women (and sometimes men) who represent care and economic giving/assistance (Malkki 2015). Celebrity advertising is another strategy for aid/development organizations to fundraise or call attention to their work. Celebrities, who self-identify as global humanitarians or supporters of development, represent the antithesis of the needy other by way of their performances of excessive wealth, conspicuous consumption and capital accumulation. Television, movie and music celebrities' "star power" adds a layer of shine and legitimacy to the projects and programs they support. The irony of their advocacy for humanitarian or development causes can be seen through their sponsorship and advertising of goods and services that both directly and indirectly create or reinforce wealth disparities, i.e., products made in factories that pay low wages and require long hours with unregulated/unsafe work environments for staff.

Therefore, it is somewhat surprising that celebrities have become the embodiment of morality by attempting to suggest what audiences should be feeling in response to particular representations of those on the receiving end of assistance (Richey 2016). The celebrity body represents a particular aesthetic, authority and ability to influence individuals. Celebrities are advocates and representatives for the selling of commodities, ideas and actions, while their likeness (image) is also a **commodity** that can be bought and sold. The celebrity as commodity continually uses his/her body to sell an array of products (Fluri 2017). Therefore, many celebrities seek to create their own "brand" to remain relevant and influential to their fanbase. For example, Louis Vuitton's Core Values advertising campaign enlisted Bono and Angelina Jolie (actress and self-identified humanitarian advocate) to pose for this campaign. In separate advertisements, Bono and Jolie are pictured with their respective Louis Vuitton bags and with written copy that highlights their advocacy for humanitarian and development causes (see Fluri 2017). This exemplifies the inextricable link between celebrity as commodity, celebrity selling a commodity and the selling of the celebrity's aid/development branding.

Humanitarian and development organizations enlist the assistance of celebrities as advertisers and advocates for humanitarian or development assistance, and this subsequently becomes part of an individual celebrity's self-branding. Celebrity humanitarianism and development branding exemplifies what Goodman and Barnes (2011) identify as "star/poverty space." They write:

> These spaces in turn make and "make up" the development celebrity at the same time that they give them their elevated and authoritative voice that draws us in and allows them to pronounce on (under)development and humanitarian crises. These images and words of development celebrities provide the basis around which charity campaigns and events are now created, popularized and marketed in order to facilitate the transnational relations of care between audiences, consumers and contributors and those "in need". Celebrities are now the cultural intermediaries, along with NGOs and charities, who encourage us to care about others, other environments and other places.
>
> (Goodman and Barnes 2011, 79)

Humanitarian/development celebrity branding calls specific attention to the ways in which gender, race and socioeconomic class backgrounds are used to legitimize a celebrity's position as a moral spokesperson for distant others. For example, Angelina Jolie went from "being decadent, unstable and sexually perverse, [to transform] into a figure of humility, propriety and sexual modesty" (Repo and Yrjölä 2011, 49). Her international fame and interests in humanitarianism have included being a goodwill ambassador to the UN 2001–12, and in April 2012 she was appointed as a special envoy to the UN. Male celebrities such as Bono and Bob Geldof (lead-singer of the Boomtown Rats and founder of Band-Aid) strengthen their claims of legitimacy to represent causes in Africa by making reference to their working/lower-middle-class and postcolonial Irish backgrounds. Bono's origin story tells of how he "became a man" in the absence of female caregiving after he lost his mother in his youth and was subsequently raised in a harsh male environment, leading him to seek out religion and music (Repo and Yrjölä 2011, 49). Discussing a celebrity's past or portraying them as contemporary exemplars of moral behavior increases their effectiveness as spokespersons of international assistance. Supporters of celebrity action and intervention suggest that celebrities fill the gaps in existing forms of humanitarianism, while detractors argue that celebrity "star power" renders invisible the more insidious forms of structural violence and material inequalities that cause food scarcities, conflicts and other disasters. Structural violence refers to institutional forms of sexism, racism, classism, etc., that marginalize certain individuals based largely on these social identities and therefore produce implicit forms of violence against those persons (Galtung 1969). For example, structural violence occurs when socially or economically marginalized people cannot access the necessary food, shelter, clothing or healthcare.

Celebrity work in many ways garners more international interest and attention to important issues; celebrities remain walking billboards of global inequalities. The extensive entourages necessary for a celebrity to go to a place reeling from disaster or conflict for that all-important "photo opportunity" further highlights the spatial, situational and economic divisions between them and the "needy other" for whom they purport to represent, assist or speak. Similar to other individuals who have amassed wealth, their daily operations rely on and profit from global capitalism and wealth inequalities. For example, celebrity aid/development branding is advanced by their involvement in international assistance or charitable giving, adding legitimacy to a "post-democratic" political environment that is controlled by elites who are not held accountable for their actions (Kapoor 2013). Moyo's (2009) critique of aid in Africa calls into question Western-oriented assistance and celebrity advocacy by arguing that these policies interfere with indigenous methods for building self-sufficient local economies and political structures. Thus, while celebrities offer an opportunity to popularize and legitimize global aid and development, they remain dependent upon and active participants in the production of economic inequality (Fluri 2017).

Drawing on capitalist consumption methods to donate or support aid and development does little to address global inequalities. These consumer-based models move the responsibility for improving wages, working conditions and access to resources onto consumers and NGOs rather than corporations and governments that are causing and perpetuating these inequalities. In addition to online shopping platforms for assistance, and organizations and celebrities "selling" their brand through charitable campaigns, conscientious consumption has become another method for consuming "new" ideas for development and capitalist-based globalization. Alternative forms of consumerism and lending have been developed in order to provide more favorable opportunities for men and women in situations of economic privation. The following case studies examine **microcredit/finance** and **fair trade**, both of which exemplify relatively recent trends that attempt to work within capitalist frameworks while attending to or assisting laborers and poor individuals by providing access to credit and "fair" wages, respectively.

Case studies: microcredit and fair trade

Microcredit/finance

Microcredit and microfinance have been lauded for providing small loans, mostly to women, in an effort to provide them access to credit, and by extension improve their lives and livelihoods. The idea of microloans began in the 1980s in Bangladesh by way of Muhammad Yunus, founder of the Grameen Bank. Yunus sought to assist women by providing them with access to credit in order to fund small businesses or invest in the existing small businesses within their communities. These small loans were distributed to women without material collateral or credit history. The initial success of his approach began to garner international attention, culminating with Yunus and the Grameen Bank receiving the Nobel Peace Prize in 2006. International attention led many donor countries, NGOs and international banks to engage in various forms of microlending activities.

Various imitations of the Grameen Bank model of microlending did not take into consideration the needs of women and their communities. For individuals without access to material collateral, loans are structured communally. Therefore, a group of eight or ten women will take out a loan collectively. If one woman cannot pay her share of the loan payment any given week, the remaining community members must contribute her share. This was intended to formalize existing informal systems of sharing economic resources in various communities. In practice, collective borrowing has led to abusive behaviors from bank employees and disharmony within communities. For loan recipients, the structural problems with microlending include short time schedules for paying back loans, high interest rates and communal loan ownership. From the perspective of many economists, microlending continues to be viewed as a safe investment option because nearly 90 percent of loan recipients pay their debts back, with interest and on time.

The focus on women as recipients of microloans has in some geographic areas increased their power over decision making within the household (Roy 2010). In other cases, such as Nepal, high repayment rates are achieved by "loan swapping," when additional loans are taken out to pay back existing ones (Rankin 2001). In other locations, microcredit programs were expected to improve women's mobility in public space, particularly in geographic areas where women are expected to spend the majority of their time at home.

Interestingly, the economic success of women receiving microloans reduced rather than increased their mobility (Kabeer 2001). These women viewed practicing **purdah** (privacy/seclusion from public spaces) as an emblem of their socioeconomic privilege, rather than mobility. This example highlights three critiques of gender-based development models: 1) paid labor is not always a clear method for empowering women; 2) what is viewed as empowering (i.e., mobility) for some women may not be desirable for others; and 3) attempting to socially engineer gender roles/relations toward a universal concept of gender-based rights disrupts the ability to understand, respect and effectively communicate across sociocultural and economic differences.

While access to credit is indeed important for many individuals at all levels of the economy, the cyclical debt associated with microcredit (i.e., loan swapping) has been at the forefront of social science critiques. Other concerns include the harsh treatment of loan recipients from lenders. For example, in Hyderabad, India, the microlending organization SKS has been criticized for using severe tactics to ensure loan payments, including physical harassment and verbal abuse. Several individuals who could not pay back their loans committed suicide, prompting an investigation that found SKS both directly and indirectly accountable, which led to the Microfinance Institutions Regulation of Money Lending Act of 2011. This Act (signed into law in 2013) banned microfinance institutions from approaching the homes of their customers, required lenders to receive approval from the government to give a second loan to the same borrower and lengthened the repayment cycle from one week to one month.

Microlending often identifies entrepreneurialism as a method for low-income individuals to improve their situations. While this may be true in some cases, it does not work in all cases. Roy's (2010) research on microcredit/finance critiques these practices by arguing that poverty has been reconceptualized as a site of potential entrepreneurship and microfinance as a form of "ethical economics." Situating the provisioning of credit (and consequently debt) to low-income individuals as ethical has further converted these types of loans as charity, when they are actually for-profit lending. For example, Roy's (2010) research on microlenders identified the ways in which

> wage-earning labor is presented as servitude and wages as charity. It is in this way that the poor are folded into the structure of microfinance – not as laboring bodies but rather as moral subjects, as either bootstrapping entrepreneurs or as lazy encroachers.
>
> (Roy 2010, 193)

Neoliberal capitalist models of development have associated poverty with failure and laziness, rather than produced through various forms of structural violence associated with intersectional social inequalities (i.e., gender, sexuality, race, class, ability), uneven development and inequitably structured economic systems.

Microcredit/finance remains generally focused on low-income women. In the United States there are various forms of microlending which do not focus specifically on women but are structured to take advantage of the economic needs of low-income individuals. Payday loans are an example of microlending that allows individuals irrespective of their credit rating to take out a small loan, with proof of residence, identification, pay stub and sometimes a bank account. These loans are intended to be short-term (one–two weeks) and include a fee for taking out the loan. However, in many cases individuals are not able to pay back this loan within the quick turnaround period and therefore take out another loan to pay back the first loan, or slowly pay back their existing loan at exorbitantly high rates of interest. Interest rates for payday loans vary by state, with the highest rate of 662 percent in Texas, and lowest rate of 154 percent in Oregon, while the states of Montana, South Dakota, Arizona, Arkansas, Georgia, North Carolina, West Virginia, Delaware, Maryland, Pennsylvania, New Jersey, New York, Connecticut, Massachusetts, Vermont and New Hampshire have fee caps that do not allow high-interest-rate debt traps (CRL 2016). Access to credit remains essential for building businesses and purchasing expensive items such as automobiles and homes, while quick turnaround times between receiving and paying back a loan along with high interest rates make this form of lending more predatory than positive for low-income individuals.

Fair trade: consuming development

Microcredit/finance was developed as a method for providing low-income individuals with access to credit, even when they do not have material collateral (or a good credit rating). Microcredit/finance offers some useful opportunities while it is also rife with abuse and unfair, predatory lending practices. Fair trade is another method for working within existing capitalist production–consumption models in an attempt to improve the lives and livelihoods of laborers, particularly in the agricultural sector for formerly colonial commodities (e.g., coffee, chocolate, tea, bananas). As another market-based mechanism that involves profiting from the labor of others (usually the poorest), it has also been found to be deeply problematic in many contexts.

Various companies market fair trade as an *added value* to their products by stating that their goods are produced through fair, rather than free, trade models. Free or market-driven trade tactics rely on low-wage labor and lack of tariffs in order to ensure cheaper goods for massive consumption and to maximize profits. Conversely, fair trade offers wages and trade practices purporting to provide economic benefits to the laborers who produce the products, while simultaneously marketing the importance of fairness to the consumer. The concept of fairness is quantified because fair-trade goods are sold at higher prices than their "free-trade" counterparts. Fair trade is often lauded as an alternative to free market economies, while it actually operates as a neoliberal economic solution to existing trade difficulties (Nicholls and Opal 2005).

Fair trade mostly focuses on agricultural production in rural areas that "perpetuates and exacerbates gendered production structures that disproportionately disadvantage women" (Rice 2010, 47). For example, market-based agricultural production remains male-dominated, while subsistence agriculture for household sustenance is the domain of women. Interestingly, while women produce 70 percent of food consumed in spaces where subsistence agriculture is a primary food source, income-based food production, including fair-trade items, remains controlled by male leadership within fair-trade cooperatives and male-headed households (Rice 2010). Female-dominated forms of production – such as handicrafts, weaving and embroidery – have had a much more difficult time seeking fair-trade certification (Rice 2010).

Gender-equal pay standards are ubiquitous among fair-trade organizations, while most studies that include a gender analysis "conclude that more needs to be done to further both gender equity within fair trade, and gender equality within societies" (Le Mare 2008, 1935). A major issue for fair-trade organizations is the lack of knowledge or interest in understanding the interactions between fair-trade employment and women's domestic responsibilities (Le Mare 2008). Without an understanding of gendered division of labor and work (both paid and unpaid), these organizations will not be able to address gender equity and equality issues. While fair trade continues to offer a somewhat viable alternative, it does not radically change existing production–consumption models because it retains a faith in markets to produce public goods. Similar to aid and development fundraising, fair-trade organizations need to advertise and sell the "idea" of fairness in order to convince customers to spend extra funds on their production. In each of these cases, consumer and market forces are privileged over the needs, wants and livelihoods of laborers or borrowers.

Fair trade in the Dominican Republic

Banana production in the Dominican Republic is dominated by certified organic and fair trade production models. Bananas produced in the Dominican Republic are almost exclusively sold in European markets relying on former colonial trade relationships. As discussed earlier, fair trade is a market-based approach to economic development that relies on the premise that "relatively wealthy consumers in first-world countries can be convinced to pay more for something that is produced in ways that reflect their values" (Getz and Shreck 2006, 491). In many respects, banana production in the Dominican Republic is driven more

Figure 3.1 Fair trade bananas in the Dominican Republic

by market forces (i.e., demand and supply) than by a true reflection of fair-trade principles (Roquigny, Vagneron, Lescot and Loeillet 2008). For example, in order to ensure the fair-trade label, products must be certified by a third party to ensure fair-trade practices throughout the means of production. However, in the Dominican Republic, "fair trade" bananas have been sold without the participation of fair-trade-certified farmers, and these farmers did not have adequate (or any) knowledge about the tenets or expectations of fair-trade labor practices (Getz and Shreck 2006, Trauger and Murphy 2012). Farmers' and workers' lack of understanding about their rights prevents them from demanding accountability from other individuals or organizations involved in the fair-trade system (Getz and Shreck 2006, 498). Therefore, both the lack of knowledge and the production of bananas without always using certified farmers result in local-level socioeconomic inequalities (part of what fair-trade advertising promises to alleviate).

Fair-trade advertising promises a more direct relationship between consumers and producers, which it must do via labeling in a global commodity chain. In the Dominican Republic, stakeholders remain the same for both fair-trade and conventional trading systems, which produces significant obstacles for creating direct contacts between producers and consumers (Vagneron and Roquigny 2011). Small banana producers are further caught between large plantations and the market demands from influential European retailers (Vagneron and Roquigny 2011). Fair trade advocates use small farm producers to help advertise the concept of fairly traded goods, which includes the belief that "fair trade" ultimately benefits small or family-owned farms. The marketing of fair-trade products perpetuates "the small farm imaginary," which supports the belief that small farms are the best scale for ensuring fair and just labor practices (Trauger 2014, 1097). However, small farms in the Dominican Republic do not meet fair trade standards and the working conditions are often worse than on large plantations (Trauger 2014). The imagined fairness and justice associated with small farmers relies on romanticized notions of small family farms, while in reality fair-trade goods are just as dependent on market demands as conventional "free" trade, resulting in the use of illegally employed Haitian laborers (Trauger 2014). Men dominate fair-trade production in the Dominican Republic; few women own or operate farms. The work is gender-segregated within the farm and women frequently work in the lower-paid jobs such as packing items for transport. Ninety percent of all workers in bananas, fair trade or otherwise, are Haitian migrants, and are vulnerable to deportation, wage theft and lax enforcement of standards, which the fair trade industry seems ill-equipped to address.

Future challenges

The business of aid and development is a multi-billion-dollar industry fraught with various levels of abuse, waste of resources, the objectification of impoverished places and peoples, and divisions between zones of accumulation and spaces of dispossession. While the merits of globalization and neoliberal capitalism are regularly touted through mainstream media sources and humanitarian and development celebrities, the division between the wealthiest and the poorest continues to grow. For example, an Oxfam report identifies:

> Since 2015, the richest one percent owned more wealth than the rest of the planet. Eight men own the same wealth as the poorest half of the world (3.6 billion people).... Over the last thirty years the growth in the incomes of the bottom fifty percent has been zero, whereas the incomes of the top one percent have grown three hundred percent.
>
> (Hardoon 2017, 2)

Celebrities help to sell the concept of assistance for the poor while simultaneously representing wealth as aspiration rather than engendering global economic inequalities. Therefore, it is important to recognize the seductions of neoliberal aid and development as part of rather than separate from economic disparities (Fluri 2017). Growing inequalities require acute and serious attention. If inequalities are not adequately addressed, the economic development business will continue to grow while not adequately or effectively improving the lives and livelihoods of the people it purports to assist. Nongovernmental development organizations are interested in tackling these inequalities, but the same or similar policies and programs of the past ten to twenty years are clearly not working. Part of the problem rests with the general public's unrelenting belief in aid and development as a viable, necessary and essential form of assistance.

As this chapter has shown, while merits can be found in various programs, much of the funding and allocation of resources continues to benefit individuals from spaces of economic privilege rather than people living in situations of economic privation. Therefore, it is imperative to highlight paradigms, programs and projects that counter neoliberal capitalism and ideas engendered within communities rather than outside of them. Consumer-based methods for ensuring labor rights can work; however, they require constant vigilance and critical inquiries. International labor rights should be paramount to profits and price points that benefit consumers while meeting the wealth aspirations of stockholders and corporate boards. A radical shift in global inequalities is necessary if not imminent, which requires acknowledging, addressing and changing the business of aid and development.

Recommended reading

Microcredit and its discontents: women in debt in Bangladesh, Lamia Karim; *Give a man a fish*, James Ferguson; *No logo*, Naomi Klein; *Brand aid: shopping well to save the world*, Lisa Ann Richey and Stefano Ponte

Recommended viewing

Empire: superclass (www.youtube.com/watch?v=aZjr6e0J0Ls); *The trouble with aid* (https://vimeo.com/81133030); *China Blue*; *Company Town*; *Africa for*

Norway (www.youtube.com/watch?v=oJLqyuxm96kRadi-Aid)

Questions for discussion

What would non-market-based assistance look like? How could celebrities use their influence and authority to improve labor rights and address economic inequality caused by corporations?

References

Bhabha, H.K. (1994). *The location of culture*. London: Routledge.

Brockington, D. (2014). *Celebrity advocacy and international development*. New York and London: Routledge.

Butler, J. (1990). *Gender trouble: feminism and the subversion of Identity*. New York, NY: Routledge.

Collier, P. (2007). *The bottom billion: why the poorest countries are failing and what can be done about it*. Oxford and New York: Oxford University Press.

CRL (2016). Center for Responsible Lending map of US payday interest rates. www.responsiblelending.org/research-publication/map-us-payday-interest-rates. Accessed 2/19/2013.

Elliott, L. (2017). World's eight richest people have same wealth as poorest 50%. *Guardian*, January 16. www.theguardian.com/global-development/2017/jan/16/worlds-eight-richest-people-have-same-wealth-as-poorest-50. Accessed 2/17/2019.

Fluri, J. (2009). "Foreign passports only": geographies of (post) conflict work in Kabul, Afghanistan. *Annals of the Association of American Geographers*, 99(5), 986–94.

Fluri, J. (2017). Crisis and consumption: "saving" the poor and the seductions of capitalism. *Humanities*, 6(2), 36. https://doi.org/10.3390/h6020036.

Fluri, J.L., & Lehr, R. (2017). *The carpetbaggers of Kabul and other American-Afghan entanglements: intimate development and the currency of gender and grief*. Athens, GA: University of Georgia Press.

Fraser, N. (2013). *Fortunes of feminism: from state-managed capitalism to neoliberal crisis*. New York, NY: Verso Books.

Galtung, J. (1969) Violence, peace and peace research. *Journal of Peace Research*, 6(3), 167–91.

Getz, C., & Shreck, A. (2006). What organic and fair trade labels do not tell us: towards a place-based understanding of certification. *International Journal of Consumer Studies*, 30(5), 490–501.

Goodman, M.K., & Barnes, C. (2011). Star/poverty space: the making of the "development celebrity." *Celebrity Studies*, 2(1), 69–85. https://doi.org/10.1080/19392397.2011.544164.

Hardoon, D. (2017). *An economy of the 99%: it's time to build a human economy that benefits everyone, not just the privileged few* (Oxfam Briefing Paper). Oxford: Oxfam International.

Hasian, M.A. Jr. (2016). *Humanitarian aid and the impoverished rhetoric of celebrity advocacy*. New York, NY: Peter Lang Publishing.

Kabeer, N. (1994). *Reversed realities: gender hierarchies in development thought*. London: Verso.

Kabeer, N. (2001) Conflicts over credit: re-evaluating the empowerment potential of loans to women in rural Bangladesh. *World Development*, 29(1), 63–84.

Kabeer, N. (2005) Gender equality and women's empowerment: a critical analysis of the third Millennium Development Goal. *Gender & Development*, 13(1), 13–24.

Kapoor, I. (2013). *Celebrity humanitarianism: ideology of global charity*. New York and London: Routledge.

Kothari, U. (2005). Authority and expertise: the professionalisation of international development and the ordering of dissent. *Antipode*, 37(3), 425–46.

Le Mare, A. (2008). The impact of fair trade on social and economic development: a review of the literature. *Geography Compass*, 2(6), 1922–42.

Malkki, L.H. (2015). *The need to help: the domestic arts of international humanitarianism*. Durham, NC: Duke University Press.

Moghadam, V.M. (2002). Islamic feminism and its discontents: toward a resolution of the debate. *Signs*, 27(4), 1135–71.

Moyo, D. (2009). *Dead aid: why aid is not working and how there is a better way for Africa*. New York, NY: Farrar, Straus, and Giroux.

Nelson, L. (1999). Bodies (and spaces) do matter: the limits of performativity. *Gender, Place & Culture Feminist Geography*, 6(4), 331–53.

Nicholls, A., & Opal, C. (2005). *Fair trade: market-driven ethical consumption*. Thousand Oaks, CA: Sage.

Rankin, K.N. (2001). Governing development: neoliberalism, microcredit, and rational economic woman. *Economy and Society*, 30(1), 18–37. https://doi.org/10.1080/03085140020019070.

Repo, J., & Yrjölä, R. (2011). The gender politics of celebrity humanitarianism in Africa. *International Feminist Journal of Politics*, 13(1), 44–62. https://doi.org/10.1080/14616742.2011.534661.

Rice, J.S. (2010). Free trade, fair trade and gender inequality in less developed countries. *Sustainable Development*, 18(1), 42–50.

Richey, L.A. (Ed.) (2016). *Celebrity humanitarianism and North–South relations: politics, place and power*. New York and London: Routledge.

Richey, L.A., & Ponte, S. (2011). *Brand aid: shopping well to save the world*. Minneapolis, MN: University of Minnesota Press.

Roberts, S.M. (2014). Development capital: USAID and the rise of development contractors. *Annals of the Association of American Geographers*, 104(5), 1030–51.

Roquigny, S., Vagneron, I., Lescot, T., & Loeillet, D. (2008). Making the rich richer? Value distribution in the conventional, organic and fair trade banana chains of the Dominican Republic. 3rd Fair Trade International Symposium – FTIS 2008 – Montpellier, France, May 14–16.

Rothenberg, P.S. (2008). *White privilege*. New York, NY: Macmillan.

Roy, A. (2010). *Poverty capital: microfinance and the making of development*. New York and London: Routledge.

Schaaf, R. (2013). *Development organizations*. New York and London: Routledge.

Trauger, A. (2014) Is bigger better? The small farm imaginary and fair trade banana production in the Dominican Republic. *Annals of the Association of American Geographers*, 104(5), 1082–100.

Trauger, A., & Murphy, A. (2012). On the moral equivalence of global commodities: placing the production and consumption of organic bananas. *International Journal of Sociology of Agriculture & Food*, 20(2), 197–217.

Vagneron, I., & Roquigny, S. (2011). Value distribution in conventional, organic and fair trade banana chains in the Dominican Republic. *Canadian Journal of Development Studies/Revue Canadienne D'études Du Développement*, 32(3), 324–38. https://doi.org/10.1080/02255189.2011.622619.

Part II

Processes in development

4 Development as dispossession

Introduction

In early 2016 construction was approved for Energy Transfer Partners' Dakota Access Pipeline (DAPL) across disputed territory of the Standing Rock Sioux Reservation in North Dakota, USA. The planned route was to run from the Bakken oil fields in the western part of the state crossing under both the Mississippi and Missouri rivers and Lake Oahe, threatening indigenous water supplies. Another path of the pipeline had previously been proposed to cross the Missouri river ten miles north of the city of Bismarck, but residents (90 percent white) opposed this route because of its proximity to their water sources. The Standing Rock Sioux initiated similar protests to ensure the protection of their water resources, which gained intensity over the course of a year. Efforts to disperse the protesters by local and federal officials included the use of water cannons in sub-zero temperatures. The conflict culminated in the denial of an easement by the US Army Corps of Engineers effectively halting pipeline construction in December of 2016, although the pipeline was approved and completed in 2017. The Standing Rock Sioux, led by elder LaDonna Brave Bull Allard, spoke for many regarding the threats that pipelines and oil spills pose for the integrity of water supplies, as well as indigenous concerns about the threats to native sacred spaces. The Water Protectors (as they are known) are led and sustained by women from indigenous communities. They defend their actions and the land stating, "Water is life."

In this case, the land and water vital to the indigenous populations in the region are seen as disposable resources for the extractive industries of multinational corporations. They are to be used in the interests of capital to advance economic development, without regard for the risks this poses to the most economically vulnerable populations. This takes place in the context of hundreds of years of illegal annexing of indigenous land for the purposes of developing and sustaining the American economy, to the detriment of indigenous health, livelihoods and well-being. The pipeline fight is a recent example, but this ranges from the destruction of food sources (bison) to the pollution following uranium mining on reservations. From this standpoint, development operates through and because of **dispossession** – or the enclosure and seizure of collectively held resources. Harvey (1990) refers to this as "accumulation by dispossession" – a process where the public is divested of resources and wealth, such as land or water, which are then *commodified* and concentrated in the hands of a few private individuals or corporations. This is facilitated through enclosure, privatization, **financialization** (or the transformation of material assets to monetary ones) and redistribution. Leadership by indigenous women against a white male-dominated industry signals a dramatic difference in values and the use of power to resist or facilitate the process of development.

Agricultural and resource extraction form the basis for most forms of economic development. State

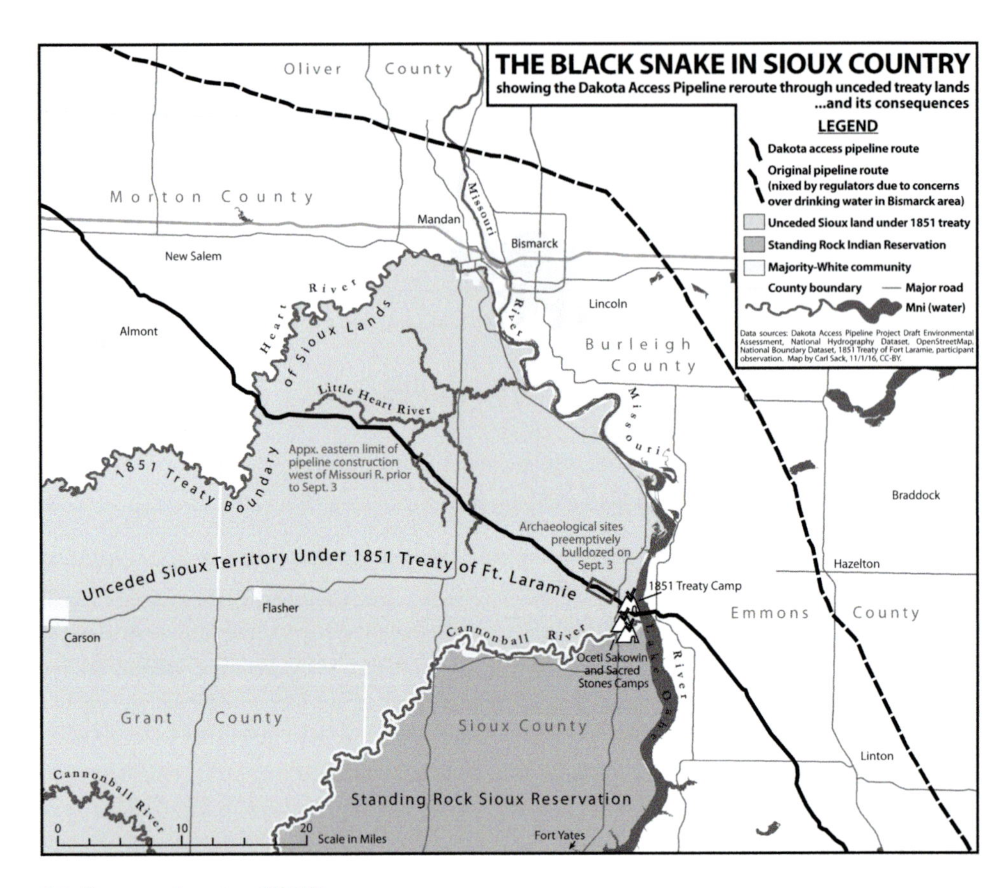

Figure 4.1 Proposed route of DAPL

or imperial acquisition of territory usually proceeds from the seizure of land from indigenous people inhabiting the land prior to colonization: the Taino and Carib in the Caribbean, thousands of tribes in North and South America, Australian aboriginal populations, Palestinians, South African tribes – the list goes on. Violent resistance almost always ensues, and those enacting violence frame indigenous populations as enemy combatants, insurgents or terrorists. If the colonizing power overwhelms the resistance with military power, treaties are often coercively forged, signed and broken to guarantee access to land, resources and sometimes slave labor. Those indigenous people unwilling to conform to the settler state's political and economic agenda (along with many who are willing) are relegated to the open-air prisons of reservations/reserves/townships/occupied territories, grinding poverty, racism and violence. At the heart of this project is the violent dispossession, via genocide to obtain illegal access to capital, the means of production, resources and territory (Coulthard 2014).

Once a settler state forms politically, it then organizes itself economically to feed its population. Many former colonies were created simply for the purposes of providing an important agricultural commodity for the metropole (Sri Lanka for tea, Windward Islands for sugar). In the postindependence era, many of these former colonies maintained trade agreements that limited their economic development to agricultural products. This arrangement was usually kept in place by installing puppet regimes composed of local elites (banana republics). Many former colonies remain poor, unstable, wracked by internal civil war and overly specialized as single resource economies as a result. Additionally, loans made to newly independent countries were meant to initiate a cycle of debt, from which few colonies have yet to emerge, and as a consequence of which the health and education sectors are gutted. Agricultural modernization in the former empires freed most workers from the work of producing food and they migrated into waged work in the manufacturing sector. This general pattern of agricultural modernization occurs everywhere today, beginning with the enclosure of common resources.

Enclosure, privatization and dispossession

Modernity has been characterized by the production of unequal social relations, in particular the production of racial categories through natural science, as well as the gendered division of labor in the production of public and private space. Modernity, with its emphasis on urbanization and wage labor relations, also constructs the urban–rural divide, which normalizes the countryside as the ideal site for the peasantry and food production. The power of this narrative has transformed societies everywhere, converting peasants into laborers and farmers into entrepreneurs (Gidwani 2008). In the past 60 years agricultural production in nearly every part of the world transitioned to a modern agricultural system characterized by a vertically integrated market (versus a subsistence) economy of food (Friedmann 1993). The commodification of food resulted in the vertical integration and the concentration of power in a few very large firms with national governments increasingly tailoring food regulation to the demands of agribusiness. Decision-making power about some of the most fundamental aspects of life – land, seed and food supplies – is now concentrated in the hands of governments, supranational organizations and transnational corporations. These institutions work together, largely through territorial-based government policies, such as structural adjustment programs, to continue the process of enclosure, and enroll small-scale producers in the global economy (Patel and McMichael 2009). This form of development impoverishes and makes hungry millions of people by removing them from the land and placing them into wage labor relations in the global economy.

The enclosure of common agricultural land sowed the seeds for the modernization and commoditization of agriculture. The land base supporting a pre-capitalist agriculture was traditionally a shared resource, held in the form of what we now call the "commons" (van der Ploeg 2010). Enclosure had the effect of privatizing land, as well as separating farmers from relations of interdependence on each other. The destruction of the commons as a mode of production allowed the rationalization of agriculture to proceed, one farmer at a time. According to van der Ploeg, enclosure and the subsequent industrialization of agriculture contributed to land losing its importance and relevance as ecological capital. Land, as the commons, had historically been the resource base that made agricultural production possible, and to which farmers contributed collectively as a means to build ecological capital. Land is now simply seen as the staging area for the conversion of capital into commodities, which in theory could happen anywhere.

Intersectionality, capitalism and the primary sector

The **three-sector theory** is a theory that divides economic activities into three distinctive categories: primary, secondary and tertiary (Fisher 1939). The primary sector involves the extraction of raw materials, or the production of basic subsistence needs (food, energy). The secondary sector consists of manufacturing and other industrial activities (refining, construction). The tertiary sector involves the extension of services, such as education or healthcare. Some theorists include the quaternary sector, the management and/or processing of digital information. This could also be considered a service, but it is increasingly shaping the economies of the wealthy, and those aspiring to become wealthy. Primary economic activities are central to development processes, often enroll men and women in highly gendered ways, and are increasingly threatened by climate change, political instability, loss of markets or recession.

The distribution of primary economic activity is geographically uneven. The spatial differentiation of these processes is a function of three main factors: the distribution of natural resources (arable land, mineral deposits), the market for the resource and the degree to which capitalization of the extractive industry has occurred. While a region or country may have significant deposits of a resource, if it is not economically valuable, and no capital investment has been made in its extraction, the industry will be underdeveloped. For example, columbite-tantalite, also known as coltan, is an important component of electronic devices, such as mobile phones. The Democratic Republic of Congo (DRC) is a leading exporter of the mineral, largely to China, but the extraction was underdeveloped until the market demand peaked in the early 2000s. Due to political instability in the region, most coltan is mined by hand and smuggled out of the country illegally. By contrast, Australia's production of coltan was controlled by a single multinational (now bankrupt) and was extracted from open-pit mines with machinery (Mantz 2008).

For extractive or productive primary economic activities to function well, they must be premised on territorial control of the resource in some form – through land grabs, colonialism or long-term leases. Imperial and capitalist acquisitions of territory are not disconnected, and countries often have access to land that was seized illegally. In addition, labor is required for the economic activity, and under capitalism, labor is compensated via wages. Under imperial or other illegal labor arrangements, people work under conditions of slavery (debt, trafficking, indentured servitude) as well as through the unpaid labor of women and children on farms and in other unregulated extractive industries. Extractive industries are most profitable where the population is vulnerable and the territorial control of the resource can be facilitated without purchasing it.

Thus, colonialism is an important precursor to the development (and normalization) of a profitable primary sector, even after the decolonization process is concluded. For example, many colonies were developed as a source of a valuable commodity (bananas, tea, sugar and spices). Once independent of imperial control, former colonies have few options but relying on this single sector for developing their economies. The former colonizer often negotiates trade deals with the newly independent colony to guarantee markets, but this only further locks an economy into a pattern of dependency. **Dependency theory** asserts that current or former colonies are trapped in situations where a colony has a product, such as bananas, and they are required to sell a certain amount to certain countries (quotas) and at certain prices, often at the expense of their own autonomy and prosperity (Reitsma 1982). The strengths of this theory indicate that development is not a universally achievable outcome for all, nor is the playing field level for many poor countries. Critiques of dependency theory suggest that it is based on outmoded notions of the capabilities of nations and individuals to navigate capitalist systems (Smith 2010). However it is understood, uneven development results from the imposition of conditionalities on loans, structural adjustment programs that starve the welfare sector, military interventions and trade arrangements that structure space and relationships to the disadvantage of the poorest countries (Escobar 2011). Capitalist investment often follows a single sector and market, leaving countries or colonies with few options for diversifying their economies, and leading to what is called a **single resource economy**, within which vulnerable laborers are exploited in intersectional ways. This trajectory is partly why, during the middle of the 20th century, some newly independent countries, such as Cuba and Venezuela, refused export orientations, and developed instead a domestic subsistence economy. Diplomatic and military conflicts, from which the capitalist states frequently benefit, often follow such resistance.

A single resource economy (SRE) is a country whose gross domestic product is produced by a single commodity or group of commodities. Countries with SREs are almost always involved in some kind of primary economic activity, such as extractive industries or agriculture. The degree to which economies experience commodity concentration is tightly linked to reliance on export markets, which further the dependency dynamic in a positive feedback loop. Debt is an almost inevitable outcome of the structure of this kind of economy and its relationships with other economies given the fluctuation of prices in global markets and changing regulation of trade. For example, coffee

FOCUS: SINGLE RESOURCE ECONOMY

Nauru is an exemplar of the problems associated with single resource economies and dependency on commodities for export. Nauru is an island in the South Pacific, and was once called the richest place on earth in the 1960s and 1970s. A former German colony and the smallest state in the South Pacific, Nauru gained independence in 1968. Nauru had easily accessible vast and valuable phosphate reserves, and the country profited immensely from its extraction until the reserves were depleted and the environment damaged. Corruption and mismanagement of a trust that was established to care for citizens when the reserves were depleted resulted in the government being unable to perform basic functions. In 2001 a Norwegian ship bearing refugees sought to dock at a port in Australia. It was refused entry and the refugees were ultimately transported to Nauru, where they formed the basis of Australia's "Pacific Solution" for refugees. Nauru received aid from the Australian government in exchange for hosting detention facilities for asylum seekers. Residents of the detention centers are held indefinitely in appalling conditions with no legal recourse (Mountz 2011). Refugee population management remains Nauru's current primary economic activity. As recently as 2018, New Zealand offered to take children and refugees from Nauru, but Australia refused to comply, citing concerns about the visas that refugees would receive in New Zealand that would allow them to travel to Australia (AAP, 2018).

prices swing wildly year to year, but growers cannot adjust production after planting coffee trees. Thus, good prices one year may induce farmers in an economy dependent on coffee to plant more trees, flooding the market in subsequent years and thus lowering prices. If the coffee cannot be sold, growers go into debt, and the dependence of the economy on this commodity triggers an economic crisis. When a country appeals to the IMF for loans to ease the crisis, structural adjustment policies are imposed. These free-market "reforms" deepen dependency on global trade, to the disadvantage of the indebted country. In addition, austerity measures are taken to reduce domestic spending on healthcare and education. The removal of basic kinds of welfare from citizens furthers a cycle of poverty, dispossession and desperation, triggering civil unrest, violence or out-migration. This has a destabilizing effect on the region, further exacerbating dependency, debt and dispossession.

Primary economic activities

Primary economic activities are those that involve harvesting or extractive industries. They include the three major sectors of agriculture and fisheries, mining and quarrying, and forestry. This also includes hunting, subsistence farming/gathering, water bottling, and topsoil mining, among many others. These activities form the basis for many other economic activities. For example, the mining of iron ore provides the raw material for steel manufacturing. Other activities less significant to the global economy are subsistence farming, hunting and foraging. While these are less capitalized activities, they support the capitalist sector with unwaged labor and the provision of basic needs for food. Waged and unwaged workers are enrolled in these economic activities in specifically intersectional ways. In addition, primary economic activities follow in the wake of colonial claims to territory; land grabs or other kinds of annexation of space by an independent country or **national state**. In what follows we briefly describe the three main sectors before we discuss their relationship to intersectional identities and processes of dispossession.

Agriculture

Agriculture is the collective sum of the processes involved in growing food for human and animal consumption. Agriculture has profoundly shaped the past 10,000 years of human activity on earth, and is a driver of economic development, urbanization and migration. It is a global industry composed of various kinds of production practices. Agriculture is found wherever there is adequate water, fertile soils, appropriate topography and long enough growing seasons. Agricultural activities include nomadic herding, shifting cultivation, intensive, extensive and urban subsistence, industrial-commercial and sustainable-organic. The primary difference between subsistence farming and commercial farming is the degree to which capitalism has converted the natural resources of soil, seeds and water into commodities for sale in global markets. The gender differences between these two systems are striking and mirror larger-scale patterns in economic activity. The smaller the profit and lower the capitalization, the more women tend to be involved. Children and slaves are widely used in agriculture in every economy in the world. It is estimated that agriculture is the largest user of child and slave labor compared with other economic sectors.

The distribution of agricultural activities over the earth's surface is highly uneven. The UN estimates that more than a third of the world's land is put toward growing food. The percentage of people involved in agriculture varies, however, by the mode of production. Intensive production that includes machinery, chemicals and high levels of capitalization involve a small fraction of the total population. Conversely, the more labor-intensive the production, the more people are required to be involved as laborers. In some countries, such as Bhutan, the proportion of people involved in agriculture is as high as 90 percent. As agricultural practices **modernize** – meaning they include new forms of technology and machinery – the proportion of people working in farming declines. Today a handful of major corporations dominate agriculture (including Monsanto and Cargill) and work in tandem with national states to draft policies that further concentrate power and capital in private hands. This has profound implications for the food systems and there are a variety of responses to this concentration of power, described in more detail below.

Fisheries

Like agriculture, fisheries can be highly capitalized, large-scale commercial operations or can be small-scale subsistence activities. Subsistence fishing is often a village or community-scale activity using handmade materials and involving a relatively small catch. It often supplements other activities such as agriculture or waged work. Commercial fishing can involve high-value line-caught fish or large-scale aquaculture in fish

FOCUS: GREEN REVOLUTION

The Green Revolution was an international effort to modernize agriculture led by the Rockefeller and Ford foundations, national state governments (India, Pakistan, Mexico and the Philippines), international aid organizations and plant geneticists during the 1960s. The widespread fear that large, growing and hungry populations would either revolt or turn to communism during the Cold War prompted these agencies and institutions to identify and mitigate the causes of hunger. They found evidence for their ideas about the relationship between hunger and the unproductivity of agriculture in the integrated crop and livestock systems of Third-World agriculture. The idea that low productivity led to hunger (and not lack of infrastructure, lack of healthcare/maternal care, pre-existing inequalities or foreign policy) led to the development of high-yielding varieties in the laboratories of Western scientists. The adoption of these varieties necessitated the introduction of monoculture, the disruption of resilient crop–livestock systems and the consolidation of land. While the Green Revolution may have produced more food calories, it did nothing to stop population growth and rising inequality, and may in fact have contributed to the intensification of both (Shiva 1991).

farms. While fish make up a relatively small proportion of the protein intake in the world's diet, millions of people around the world rely on fishing for subsistence or livelihoods. The general trend that the higher the capitalization, the more men are involved as laborers and decision makers holds for fishing as well as agriculture. Women are often involved in subsistence-scale fishing, and procuring food for households using handmade materials. Men are more likely to work on commercial fishing rigs and to be involved in industrial-scale fish farms.

Forestry

Forestry and timber industries seek to turn the renewable resource of plants, often in the form of trees, into industrial products, secondary products such as wood pulp and consumables (paper, furniture, cosmetics, chopsticks). Forestry takes place in every appropriate ecological zone, and is carried out in a variety of ways. Practices can include the clear-cutting of old-growth forests with valuable hardwoods such as in Brazil's rainforests. Another practice is the community-scale forestry sponsored by multinational corporations, such as the Body Shop's investment in the production of shea butter in Mali. There are other more common practices such as the cultivation of trees around villages for specific medicinal purposes, such as in rural India. Like agriculture, forestry is practiced by people with varying degrees of capitalization and sustainability, and can be undertaken on the scale of a backyard to constant harvesting of thousands of acres of trees. Both men and women work in forestry, and like other primary economic activities, the more heavily capitalized and industrialized the activity, the more likely men are to be involved. Women are often involved in tending village or community-scale forests, while men are more likely to be employed in harvesting timber for large-scale corporations.

Figure 4.2 Rural land use in the Dominican Republic

Mining

Mining and quarrying are primary activities dedicated to the extraction of minerals, fuels and other industrial raw materials from the earth. The earliest accounts of mining are from the Pàleolithic, some 40,000 years ago in Southern Africa. Today mining is a multi-billion-dollar industry dominated by **multinational corporations**. The most significant extractive industry in terms of scope and value is fossil fuels. Other industries include uranium mining or industrial material extraction for electronics, such as coltan. According to the World Bank (2017), non-renewable resource extraction is the primary economic activity for the world's least developed countries, leaving them dependent on global markets and vulnerable to geopolitical and economic volatility. Extractive industries tend to be very polluting, and require strong state oversight to prevent contamination. Most mining occurs in the context of transnational companies mining land in the spaces of dispossession, exposing and exploiting vulnerable populations in intersectional ways, such as those we elaborated on in the opening vignette. Most miners are men, but as we discuss below, women are an important part of artisanal mining, which involves new and more environmentally sensitive practices.

Development themes in agriculture, forestry and mining

The capitalist development model enriches the world's wealthiest white males at the expense of the world's working poor, and produces unprecedented levels of inequality. In 2016, the eight richest people had more wealth than 50 percent of the world's population. All but two of those people were white, American men. One of the richest men in the world, Bill Gates, makes much of his money by promoting development projects that replicate and repeat the patterns of dispossession and intersectional exploitation that we discuss above. Development does not always lift people out of poverty and into the middle class. It perpetuates a status quo in which the rich profit immensely from capitalism. In what follows, we discuss how dispossession and exploitation function in the world's largest primary sectors by examining three themes of intersectionality with respect to capitalism and the primary sector: gendered access to land; men and mining; and forestry and gendered livelihoods.

Women and gendered access to land

Feminist critiques of agriculture have revealed how women are marginalized from decision-making roles, and how gendered space plays a role in marginalizing women from access to power and knowledge in a variety of contexts, including agriculture (Whatmore 1991). An estimated 80 percent of the world's farmers are women, but their role in food production is often overshadowed by the more charismatic activities of male decision makers regarding trade and food policy. Women have always held key roles in food production from the smallest peasant household to the largest contract farm enterprise, but they often participate in unpaid and **socially reproductive labor** that contributes to the maintenance of household and families, and is undercounted and often undervalued in formal economic assessments of agricultural productivity. Migration from rural areas to urban areas exacerbates this pattern, because women assume men's work on farms when they leave to work in cities. This often leaves women in control of and responsible for the labor on farms they may not legally own.

The process of marginalizing women from land ownership is part of the process of colonialism. Colonization by Europeans throughout much of the world privatized collectively held land to facilitate the sale of land to individual settlers. Under colonial law in most places, only men were allowed to purchase land or hold title to it. After independence, these cultural and economic patterns persisted. While women perform much of the work in agriculture, they often lack: 1) legal standing with regard to the land, usually due to patriarchal household dynamics and inheritance laws; and 2) recognition for the importance of gendered ways of knowing about the environment. The undervaluing of women's roles and work and the ongoing process of dispossession makes them vulnerable to economic, political and environmental change in a variety of ways. The Millennium Development Goals aimed to give women greater access to property and inheritance rights, but that effort has fallen far short of its promises.

Indigenous women, if they retained access to land after colonization, must go through a highly racialized political process to gain legal standing to claim the land. Traditional claims to land are not recognized by most states, and indigenous women have the additional battle with how the state often differentiates between men and women in terms of legal status. Access to land is key for ensuring food security, and women struggle to have the same legal access to land as men. Women in Bee's (2013) study area in Mexico, for example, controlled only 13.5 percent of the arable land, while men controlled 57 percent. Up to a third of women in the community also did not have decision-making power over land they could access, or were landless. Radcliffe (2014) writes that indigenous women in Ecuador struggle to produce food on small, marginalized pieces of land, while being refused access to legal title to their land, and simultaneously advocating for more traditional notions of communal management of resources.

Nonindigenous women face similar barriers to accessing land, and risks associated with landlessness. In India, for example, Agarwal (2003) writes that the lack of access to land is a key source of vulnerability to poverty. Accessing and using land is an indicator for improved child and women's welfare, and even a small plot of land can significantly reduce poverty and food insecurity. Like Bee's study in Mexico, few women have a legal relationship to land they work in India and the political will to change this situation is virtually absent. At best, women are given title to land previously in their husband's name in some states only if they are married. Naybor (2015) also found that the lack of access to land for Ugandan women was a key source of their vulnerability, and legal ownership of land reduced dependence on foreign aid and assistance. Women in the region are collectively organizing for land reform as well as recognition and enforcement of customary rights to land. We discuss the way women challenge gender norms in more detail below.

Mining and masculinity

Access to mineral resources follows the same general process of state/colonial acquisition and dispossession as agriculture. Like agriculture, mining has long been associated with men and masculinity and women's work within the mines has been devalued. Historically, the work of mining has been gender-segregated and geographically differentiated. Men dug the mineral or ore, while women carried and processed it. As in most industries, women are paid less and experience fewer workplace protections. However, all laborers are negatively affected by unregulated working conditions in mines. Examples range from black lung disease among men in the Appalachian coal mines of the United States (Scott 2007), to the sexual harassment, rape and sex trafficking of women working in manganese mines in South Africa (Lahiri-Dutt 2012). Children continue to be significant workers in mines, with an estimated one million children working in mining (Lahiri-Dutt 2012). They often work without pay, as do many adults through debt slavery. Many countries, but not all, now ban the use of child labor in mining. Technological change negatively affects workers, who often lose their jobs when more efficient and less expensive machinery replaces the labor of human bodies.

According to Scott (2007), the masculinity of mining men in Appalachia is "dependent" on high-paying jobs available in the mining sector. Loss of these jobs is seen as a threat to both the subjectivity of men and their **social personhood**, or worth and status as an adult human, in the community. Mine work is the result of the exchange of one's labor for a wage, granting a form of citizenship and belonging to men who did not own property but only their own labor. The acceptance of mechanization and the replacement of human labor are tied to a "gendered understanding of work, embodied in the heterosexual white male breadwinner, [which] gives shape to a specific configuration of masculinity that gains moral worth from family-wage employment" (Scott 2007, 486). This masculinity is linked to notions of independence and economic rationality, tightly associated with the ownership of land and the capacity for self-determination. Mine work is central to understandings of self-worth for

Appalachian men who see providing for their families as the only kind of legitimate work for an independent head of a household. This notion of masculinity constructs itself in opposition to women and nonwhite men, who are largely excluded from mine work, but is paradoxically "dependent on the availability of high paying jobs that have been generated by the inequalities of the wage structure and labor market" (Scott 2007, 493).

Women are key to the construction of notions of masculinity because gender is a relational concept; masculinity is premised on not being defined through the binary construction of feminine attributes (i.e., strong vs weak; hard vs soft). In spite of the widespread belief that miners are men, women have worked in mines for centuries in every part of the world (Lahiri-Dutt 2012). The work of women and children in mines has been undervalued compared to men's and they were the first to lose their jobs to mechanization in the 20th century. Hegemonic notions of masculine qualities such as hardworking, strong, brave and tough enough to take on the dangerous and dirty work of mining prevail, despite the presence of women workers who are equally capable of operating machinery. Lahiri-Dutt reports that the labor of men, women and children was replaced by machines in Bolivia, Indonesia and India, leaving only men to operate the machines in the only legitimate form of mining labor. Below, we discuss in more detail the need for women to reclaim space in the mining economy, and how that might be accomplished in the DRC.

Forestry, gendered livelihoods and resistance

Forests are essential resources for every kind of economy in the world. They are key natural resources for multinationals as well as vital to the local resource needs for villages and peasants in some of the poorest regions of the world. Historic battles for control over forest resource harvesting have framed the debate around use, conservation and preservation since the mid-20th century. For example, the Chipko movement that began in the Garhwal region of the Indian Himalayas in the 1970s launched international movements to protect forests from overharvesting. In 1973, the Indian state sought to restrict collective village use of forests, and to allow the sale of harvesting rights to external companies in a sensitive region. This would have negatively impacted villagers dependent on forest resources for food, medicine and building materials. After the harvesting of valued timber in one region, women and girls surrounded the trees, dubbing themselves "tree huggers" or *chipko* in the local language, to prevent the further harvesting of trees. Their resistance sparked an international movement focused on protecting and preserving forests for the indigenous, tribal and peasant populations in nonviolent ways. The women who led the movement were its mainstay during both the resistance and the reforestation projects that followed. They were awarded the Right Livelihood Award, sometimes called the alternative Nobel Prize, in 1987 by Sweden for their efforts.

Forests are not just natural phenomena; they are social creations, often consisting of complex interactions between people, animals and crops. They are key sites of biodiversity and livelihoods for the world's poorest, and are often "distinct forest ecosystems including agroforests" that are composed of a wide variety of living and nonliving features (Rocheleau et al. 2001, 4). For example, the Zambra-Caucey agroforest in the central Dominican Republic includes houses, trees, farms, gardens, livestock and wildlife, both native and imported species, and is home to 12,000 people. This agroforest is managed by the residents in complex ways, and confronts the competing claims for development and conservation. According to Rocheleau et al., decision making is a gendered and classed process in this forest, due to the existence of power relations within households regarding land use and production decisions, and key differences in decision making between large and smallhold farms. They found that external global actors influenced the polarization of gender and class differences, with the greatest power and decision-making authority shifting over time from women smallholders to men on large farms.

When decision-making power over forests shifts from local control to multinational or transnational control (e.g., in a land grab or the leasing of resource rights from the state to an multinational company), indigenous and peasant activists, often led by women, resist the change, largely because of the threats to their livelihoods. In Ecuador, for example, women disproportionately suffer from the loss of land to external forces (the state, multinational corporations), and their struggles to retain access to (but not necessarily ownership of) land is largely invisible and ignored. In Honduras, Berta Cáceres, an indigenous leader and activist, cofounded the Council of Popular and Indigenous Organizations of Honduras (COPINH), which was formed to protect indigenous rights and protest illegal logging, and other forms of development on indigenous Lenca land. In 2006, she led a protest against the illegal construction of a hydroelectric project, which would have destroyed forests and the livelihoods of the Lenca people (McSweeney and Pearson 2013). After a long and sometimes violent campaign of resistance, Cáceres was placed under house arrest and ultimately murdered in 2016 by men suspected of having links to US military intelligence.

Case studies: the United States and the Democratic Republic of Congo

In what follows we present two examples of alternative forms of development occurring in the context of state- and market-led pressures to develop primary economic activities such as agriculture and extractive industries. The first example describes women farmers in the United States in their efforts to educate themselves and each other about sustainability and food production in the context of masculinist work cultures and patriarchal land control. The second case study examines women and artisanal mining in the DRC. The dominant narrative (and most common context) about women in mining professions is one of victimization and exploitation by men – both mine owners and fellow workers. This work shows how women subvert this narrative and engage in alternative forms of mineral extraction that is more suited to their needs.

The United States

In the context of the United States, agriculture proceeded from indigenous land grabs. The theft of land by settlers was justified on paper through a variety of mechanisms including treaties ceding land to governments, private sales of land and the Homestead Act. Agriculture took a variety of different forms throughout settler states, with plantation agriculture controlled by elite families in the South, grazing of livestock on "public" land in the West and variants of "family farming" in the Midwest and Northeast. Just as sharecropping followed from the abolition of the plantation–slave system, the collapse of entrepreneurial "family farming" in the Midwest was followed by increased vertical integration of agriculture via corporate contracts. Today, the vast majority of farmers are sole proprietorships run as franchises of major corporate brands with farmers providing the labor, infrastructure and inputs in exchange for a guaranteed price for the commodity (e.g., soybeans, broilers) at the end of the cropping season/life cycle. A small number of farms are still entrepreneurial and sell their products into local, regional or specialty markets.

Regardless of the farm type or geography, family farming in the United States has traditionally been an enterprise controlled by men, in which men are assumed to be the "farmer" and women are assumed to be the "farmwife." The state-sponsored arm of education (frequently known as Extension) for farmers has historically participated in promoting the gendered division of labor both on the farm and in the home by creating programming for a class of male producers and women consumers in rural communities, which ignores and obscures women's productive work on farms. The Extension approach to education promotes entrepreneurial models in step with the globalization and industrialization of agricultural production, including, for example, the adoption of pesticide-intensive production practices and biotechnology. Dominant ideologies of agriculture have resulted in the destruction of soil and water and poisoning of the environment. Consequently and concurrently, women's contributions to the farm economy have become devalued and the production of commodities has taken precedence over non-commodity farm enterprises and home-based subsistence.

Dominant ideologies of masculinity and femininity are critical to maintaining the unequal power relations between men and women, but women are challenging these ideologies by participating in the newly emerging markets for food produced through sustainable methods for local consumption. These new spaces have the potential to be constructed as sites of resistance from which we can witness the creation of new gender identities and the sustenance of alternative and sustainable methods of food production (Trauger et al. 2010). PA-WAgN (Pennsylvania Women's Agricultural Network) is one example of outreach and education models that meet the needs of the women farmers by providing educational opportunities through **peer learning** (where farmers teach and learn together) and highly interactive pedagogical techniques. Both of these approaches have the potential to dramatically transform the relationships between educators and women farmers, as well as to provide legitimization and opportunities for empowerment through education.

The DRC

The DRC is a country rich in mineral wealth, but ravaged by war between 1998 and 2003. The Great African War started years before in the region, and continues in small-scale civil conflict to this day, in spite of it being more than a decade since a peace agreement was signed. Sexual and gender-based violence (SGBV) was widely used against women (and men) during the conflict, and also continues to be a threat to security in the region. The mining sector in the DRC is its most valuable sector, accounting for 70–80 percent of export earnings, and is especially dangerous for women and children. Industrial (mechanized) mining is in decline, and artisanal and small-scale mining (ASM) now dominates. As discussed above, discrimination faced by women in mines is a fact of life nearly everywhere in the world where there are mines. Women are paid less than men for the same work, are often excluded from decision-making roles and are often sexually harassed and assaulted.

The DRC is one of the richest countries in terms of its mineral resources. Its mineral deposits are rich and varied, and include gold, diamonds, cobalt and copper. ASM is mining by hand with rudimentary materials such as shovels and buckets, and the low barriers to entry and the high-value product mean that ASM could provide short-term economic opportunities for women who lack formal skills or training, or who are landless. ASM is a key economic activity for individuals and families, and for the state's export economy, but women must live and work under threats of violence, inequitable pay and opportunities, and dangerous and polluted working conditions that are hazardous to children. While ASM is an important economic activity for women, the sector needs reform.

Legal protections for women and the governance and regulation of the mining industry are crucial parts for the state to play in mitigating violence against women and building a sustainable peacetime economy. Toward this end, a project designed to help women transition out of mine work into other small-scale economic activities targeted the construction of masculinity – foregrounding ideas of "protection, fatherhood, responsibility and support" (Hayes and Perks 2012, 538). This project focused on men and masculinity, and urged men to recognize women for their valuable roles in society, family and the economy. The approach to empowerment of women focused not just on the removal of women from dangerous environments and the cultivation of other skills; it focused on "a broader framework for social change – and to address men, women and the relationships between them" (Hayes and Perks 2012, 538).

Future challenges

Development lies at the intersection of the government and capitalism. In this chapter we discussed primary economic activities as vital to both meeting the needs of people and citizens and answering to the economic imperatives of the national state. This occurs via the dispossession of indigenous people, the production of inequality and the creation of surplus, and the production of capital for the elite. Roy et al. (2016) argue that development is "working as it was designed to work," which means it is supposed to produce poverty and desperate classes. This proceeds in concert with the destruction of the environment and resource base, and the murder and imprisonment of those who attempt to defend it.

An intersectional gender analysis of these processes indicates that women are disproportionately affected by the dispossession of resources and that much of the unpaid labor to reproduce a population of extractive industry workers is undertaken by women. We also find that intersectionally oppressed and embodied men and women are uniquely positioned to have insight into how to make natural resource extraction and production more just, sustainable and humane, given their frequent experience of exploitation and oppression in these systems. Below we present some possibilities for people-centered development, or economic activities that provide for people, while protecting the resource base for the present and future economies while also not engaging in intersectional forms of oppression. By this we mean treating women as agents, and regarding their economic activities, which include small-scale or environmentally protective production methods, as rational.

Capitalist economic development in the primary sector triggers some of the most significant challenges facing the world today. Climate change, as a global phenomenon, affects everyone everywhere, albeit in different ways. Climate change is caused by and exacerbated by extractive industry economies such as the use of fossil fuels for power, and agricultural activities such as the use of nitrogen fertilizers. Climate change negatively affects forests by altering the environment in ways to which plant species struggle to adapt, impacting productivity. Forests and grasslands, however, are important sources of carbon sequestration for climate change mitigation, but they must be cultivated over long periods of time. Thus, climate change is intimately linked to all major primary economic activities, as well as to the people who directly depend on a sustainable resource base for their livelihoods. The only populations not at direct risk from climate change are the oligarchs who profit from the state-capitalism nexus, which is why they deny its existence. In general, they have the wealth and privilege to insure or buy their way out of environmental disasters or move to places characterized by lower risks. According to prominent climate change activists, gender justice is inextricably linked to environmental justice, and systems that benefit humans will benefit nature as well (Terry 2009).

A significant and growing challenge to the corporate control of food production, trade and regulation of the food system is a global movement operating under the banner of food sovereignty (Trauger 2017). The meanings of food sovereignty are contested, but at its heart food sovereignty is both a definitional and a material struggle. Activists for food sovereignty position themselves against the corporate food regime in order to expand the meaning of human rights to include a "right to food" as guaranteed by the United Nations Declaration on Human Rights. Food sovereignty confronts what Malik Yakini (2013) of the Detroit Black Community Food Security Network calls the "twin evils of white supremacy and capitalism." It is a struggle that emerges from the margins by and on behalf of the poor, the hungry and the landless to reclaim spaces of decision making in the global food system. The discursive battleground lies in an ambitious redefinition of the political, the socioeconomic and the ecological in the food system. This global struggle takes a variety of forms, and aims to undermine corporate control of the food system through civil disobedience, subsidiarity and new forms of democratic decision making. Many international peasant organizations are led by peasant women, and a central plank of the food sovereignty platform is the end of violence against women.

We opened this chapter with a vignette of indigenous resistance to the construction of a pipeline in North Dakota, USA. This struggle by indigenous people to resist and gain more control over the extractive process and its consequences is not an isolated incident. Countless other struggles, particularly indigenous resistance to the construction of petroleum pipelines or oil extraction on unceded territories in North and South America, characterize access to and use of natural resources. These struggles are new, complex and here to stay. For example, proposed pipelines across disputed indigenous territory in Minnesota would cross a network of sensitive lakes and rivers. An accident, which is more common than anyone would like to believe, would be catastrophic, particularly for the tribes who rely on subsistence fishing and rice harvesting. Enrolled members of the tribes in northern Minnesota have vigorously opposed the construction, including engaging in civil disobedience by rice harvesting off the reservation to provoke a legal battle over treaty rights. The legal battle could expand tribal rights off the reservation as well as frustrate construction plans. This is just one of many battles currently being waged, frequently by women who advocate from the standpoint of gendered implications of development. These women (and men) also work for a more intersectional politics that is not premised on raced, classed and gendered forms of injustice.

Recommended reading

The American way of eating, Tracie McMillan; *Stuffed and starved*, Raj Patel; *Mining in Cornwall and Devon: men and mines*, Gill Burnley; *Your money or your life*, Eric Toussaint

Recommended viewing

North Country; *We are not slaves*; *Food Inc.*; *Power of community*; *Food chains*

Questions for discussion

How and why did food become a commodity? Who produces your food? Why are certain groups (women, migrants) involved in different ways with food production? We receive water for a small fee in municipal areas through a subsidized and municipally managed system. Do you think food could be delivered in a similar way? Why or why not? Who would benefit from such a system? What minerals do you use on a regular basis? Where do they come from? Who mined them?

References

AAP (2018). Nauru rejected NZ's refugee offer. www.theaustralian.com.au/news/latest-news/nauru-rejected-nzs-refugee-offer-nz-pm/news-story/e039317597c6f87ab31f2afa76343dba. Accessed 9/18/2018.

Agarwal, B. (2003). Gender and land rights revisited: exploring new prospects via the state, family and market. *Journal of Agrarian Change*, 3(1–2), 184–224.

Bee, B. (2013). Who reaps what is sown? A feminist inquiry into climate change adaptation in two Mexican ejidos. *ACME: An International E-Journal for Critical Geographies*, 12(1).

Coulthard, G.S. (2014). *Red skins, white masks*. Minneapolis, MN: University of Minnesota Press.

Escobar, A. (2011). *Encountering development: the making and unmaking of the Third World*. Princeton, NJ: Princeton University Press.

Fisher, A. (1939). Production, primary, secondary and tertiary. *Economic Record*, 15(1), 24–38.

Friedmann, H. (1993). The political economy of food: a global crisis. *New Left Review*, 197, 29–57.

Gidwani, V. K. (2008). *Capital, interrupted: agrarian development and the politics of work in India*. Minneapolis, MN: University of Minnesota Press.

Harvey, D. (1990). *The condition of postmodernity: an enquiry into the conditions of cultural change*. Oxford: Blackwell.

Hayes, K., & Perks, R. (2012). "Women in the artisanal and small-scale mining sector of the Democratic Republic of the Congo." In P. Lujala, & S.A. Rustad (Eds) *High-value natural resources and post-conflict peacebuilding*. New York, NY: Routledge (pp. 529–44).

Lahiri-Dutt, K. (2012). Digging women: towards a new agenda for feminist critiques of mining. *Gender, Place & Culture*, 19(2), 193–212.

Mantz, J. W. (2008). Improvisational economies: coltan production in the eastern Congo. *Social Anthropology*, 16(1), 34–50.

McSweeney, K., & Pearson, Z. (2013). Prying native people from native lands: narco business in Honduras. *NACLA Report on the Americas*, 46(4), 7–12.

Mountz, A. (2011). The enforcement archipelago: detention, haunting, and asylum on islands. *Political Geography*, 30(3), 118–28.

Naybor, D. (2015). Land as fictitious commodity: the continuing evolution of women's land rights in Uganda. *Gender, Place & Culture*, 22(6), 884–900.

Patel, R., & McMichael, P. (2009). A political economy of the food riot. *Review*, 32(1), 9–35.

Radcliffe, S.A. (2014). Gendered frontiers of land control: indigenous territory, women and contests over land in Ecuador. *Gender, Place & Culture*, 21(7), 854–71.

Reitsma, H.J.A. (1982). Development geography, dependency relations, and the capitalist scapegoat. *The Professional Geographer*, 34(2), 125–30.

Rocheleau, D., Ross, L., Morrobel, J., Malaret, L., Hernandez, R., & Kominiak, T. (2001). Complex communities and emergent ecologies in the regional agroforest of Zambrana-Chacuey, Dominican Republic. *Cultural Geographies*, 8(4), 465–92.

Roy, A., Negrón-Gonzales, G., Opoku-Agyemang, K., & Talwalker, C.V. (2016). *Encountering poverty: thinking and acting in an unequal world*. Berkeley, CA: University of California Press.

Scott, R.R. (2007). Dependent masculinity and political culture in pro-mountaintop removal discourse: or, how I learned to stop worrying and love the dragline. *Feminist Studies*, 33(3), 484–509.

Shiva, V. (1991). *The violence of the Green Revolution: ecological degradation and political conflict*. London: Zed.

Smith, N. (2010). *Uneven development: nature, capital, and the production of space*. Athens, GA: University of Georgia Press.

Terry, G. (2009). No climate justice without gender justice: an overview of the issues. *Gender & Development*, 17(1), 5–18.

Trauger, A. (2017). *We want land to live: making political space for food sovereignty*. Athens, GA: University of Georgia Press.

Trauger, A., Sachs, C., Barbercheck, M., Kiernan, N.E., Brasier, K., & Schwartzberg, A. (2010). The object of extension: agricultural education and authentic farmers in Pennsylvania. *Sociologia Ruralis*, 50(2), 85–103.

van der Ploeg, J. (2010). The peasantries of the twenty-first century: the commoditisation debate revisited. *Journal of Peasant Studies*, 37(1), 1–30.

Whatmore, S. (1991). *Farming women: gender, work and family enterprise*. Houndmills: Macmillan Academic and Professional Ltd.

World Bank (2017). Extractive industries. www.worldbank.org/en/topic/extractiveindustries/overview#1. Accessed 10/13/2017.

Yakini, M. (2013). Address to Yale Food Sovereignty Conference. www.youtube.com/watch?v=_LaMt9HVQFY. Accessed 05/11/2015.

5 Labor, migration and capital accumulation

Sunil is the second son of a Hindu family living in the Maharashtra region of central India. His eldest brother Sanjay uses products of the Green Revolution – hybrid seeds, chemical pesticides, a tractor and nitrogen fertilizer – to work the family farm. The profit from the farm is barely enough to cover expenses for Sanjay's family, care for Sunil's aging parents and pay for the upkeep of the farm. Sunil's labor is not needed on the farm, and having few other options for work in his rural village, a ninth-grade education and limited skills, he migrated to work in Mumbai and to chaperone his eldest sister, who migrated to work as a maid for a middle-class family. Sunil found little in the way of work, and resorted to selling gum in the streets before he found work in construction. He and his sister live in informal housing with a relative. They lack access to running water, sanitation and electricity. After six months, the city razed the slum in order to allow developers to construct commercial buildings. Residents refused to leave, saying they were promised affordable housing when the slum was razed, but the planned homes have not materialized. When slums are cleared for construction, people lose both their housing and any investment of labor or capital they put into their homes. Women and children are the most vulnerable to this form of dispossession. Advocates for property rights assert that titles should be granted to occupants, but this is a contested idea, one fraught with several intersectional contradictions, such as titling customs that do not allow women to own property.

In keeping with our central claim that poverty is essential to the workings of capitalism-as-development, we argue that the vulnerability of workers is generated through agricultural modernization, out-migration and entry into waged work. If subsistence agricultural livelihoods were an option, some landed farmers and peasants would choose to stay in their rural communities instead of leave. In many cases, migrants lack skills and education and must work in the **informal economy**, rather than in formalized manufacturing sectors. For example, a street vendor occupies a sidewalk or a right of way to resell food purchased from a farmer at the edge of the city. People make do with the resources that they have available to them. The government does not provide, and neither do formalized markets. In different models of state-led development, the role of the state, the market or some combination of both are charged with providing the schools, housing, clinics and transportation that civil society informally provided prior to and during colonialism.

Imperialism and capitalism are regularly justified on the basis of bringing needed infrastructure, investment or social welfare to impoverished places. History suggests, however, that the construction of infrastructure has been in the interest of ensuring the economic needs of the colonizing power, rather than benefiting the wider colonized society. For example, India's world-class train system is often lauded as the hallmark of British benevolence in India, obscur-

Figure 5.1 Rural Garhwali migrant returning home for a visit

ing the way in which the train system was designed to facilitate military activities and extract natural and human resources from India's territory. This narrative of progress further obscures the legacies of colonialism and imperialism, such as the hardening of the caste system, the imprisonment of activists for independence, the destruction of existing political systems and the plundering of India's wealth.

FOCUS: MIGRATION

Migration is typically understood as any long-distance move to a new place, sometimes temporarily, sometimes permanently. Migration can be forced or voluntary, although the distinction between those two categories is often blurred. A forced migration is often undertaken in the context of trafficking, but people may leave their homes unwillingly in the wake of a natural disaster or flight from conflict. Voluntary migrations typically involve a pull factor of economic opportunity or political asylum, with the push factors of a stagnating economy, discrimination or persecution. Migrations can be domestic or international. International migrations involve the crossing of borders, or the imaginary lines that demarcate political territory. Forced international migrations (refugees) often result in concentrations of stateless people who have no home country and no rights associated with citizenship in their host country. Domestic migrations refer to people who stay within their state's boundaries, but may involve the loss (or gain) of rights, political autonomy or livelihoods. Domestic migrants are also known as **internally displaced persons**.

In this chapter we discuss the role of **agricultural modernization** in facilitating the migration of workers from rural areas into urban areas in search of waged work. They encounter development in the form of infrastructure in two ways: 1) migrants encounter opportunities to work in waged labor (often informal) in the construction industry, which then facilitates the 2) housing, workplaces and other infrastructure (i.e., warehouses, transportation) needed for formal waged work in manufacturing. This general pattern of agricultural modernization (and decline of agricultural economic viability) and development of manufacturing continues today at a global scale. Push factors such as land grabs and the industrialization of agriculture, also known as the **agrarian crisis**, create unemployment or underemployment and increase the landlessness of rural peasantry and rural out-migration (Gidwani, 2008). For example, in one of the case studies discussed below, we highlight international migration from South Asia to work in construction in the oil-rich Gulf states.

Pathways to development, revisited

As discussed in previous chapters, development usually proceeds along one of three tracks. Export orientation is the development model most favored by capitalism and is characterized by investment, rapid growth and the sale of high-value consumer products in global markets. These economies tend to receive investment in the form of foreign capital, and frequently rapid growth is quickly offset by inflation and recession in what is called the "race to the bottom" (explained below), most recently experienced by Asian countries (e.g., Singapore). Self-sufficiency is a path taken by isolated or isolationist states that seek to stimulate economic growth by providing for their citizens' basic needs for food, shelter, etc. These economies tend to be slow-growing, protectionist and highly planned as well as frequently led by autocratic leaders. Examples include North Korea, the former Yugoslavia, Cuba and Venezuela. Mixed economies integrate an export model with a centrally planned economy that seeks to meet basic needs domestically, such as European socialist democracies. In what follows we outline some basic features of each kind of development and how it relates to waged work, migration and infrastructure. Labor is enrolled in these kinds of workforces in intersectional ways, with consequences and outcomes that are differentiated for women and men. We discuss this after a brief introduction to economic development strategies.

Market economies

Waged workers in the secondary sector of market economies are often low or unskilled and usually international, rural migrants, forced out of agricultural work by modernization and mechanization. In 2016, for example, according to the US Bureau of Labor, nearly 30 percent of construction jobs were held by people, mostly men, of Hispanic and/or Latinx origin (USDOL, 2017). The twin forces of American corn exports to Mexico and the Green Revolution in Mexico propelled men who would have worked in agriculture into other kinds of work outside the boundaries of the country. The jobs in construction sectors tend to be median wage and semi-skilled, and the availability of work is often greater in a more developed or capitalist economy. Infrastructure is vital to an export-oriented model in a vast array of sectors: factory and warehouse construction, multinational headquarters and transportation networks to facilitate the manufacturing of export products and the movement of goods and products to their intended markets.

Economic development in the context of export orientation is characterized by the "ladder" model developed by Rostow in 1960. He identified five stages of economic growth, which include a traditional society (usually agricultural), preconditions for takeoff (commercialization of agriculture, infrastructure development), "takeoff" (such as urbanization, industrialization and **foreign direct investment** (FDI)). This is followed by a "drive to maturity," where economic development spreads to multiple industrial sectors, and peaking in an "age of high mass consumption" in which the industrial sector dominates, wages are high and consumption is normative and widespread. This development model reflects the ideologies of Western liberal democracies, and is not easily applied to spaces of dispossession. Scholars critique Rostow's model because it is overly simplistic, does not address social inequality within and across countries and encourages the adoption of unsustainable mass consumption as an economic strategy (Lawson 2014). Many newly developing countries never achieve the high mass consumption target for a variety of reasons, including the suppression of wages. This model is also premised on the philosophy of **market fundamentalism**, in which it is believed that free market mechanisms have the capacity to solve the world's most pressing social and economic problems, such as poverty.

In the 1960s **import substitution** (the practice of producing something domestically rather than importing it) failed as an economic development strategy for poorer nations, and export orientation was introduced through **special economic zones** (SEZs) or geographic areas set aside from the territory of the nation-state in which some laws (e.g., labor regulations) are suspended, and export limitations (e.g., taxes) and restrictions on capital investment (e.g., bans on FDI) are reduced. According to Ong (1991), host states imagined a surplus of rural male migrants – those thought to be most likely to cause political unrest from the commercialization of agriculture – would migrate to work in the newly built offshore facilities. Instead, foreign multinationals sought young single women to create a new industrial workforce where none had existed before. We discuss this new form of accumulation, called **flexible accumulation**, and its consequences in more detail below.

Free market economies are often characterized by what is called the **race to the bottom** (the process by which corporations seek to find the lowest wages to maximize profits) and through which wages are suppressed through government deregulation of businesses or taxes. This is the result of competition between geographic areas within a particular industry or labor market for the cheapest labor market and the least regulated environments. The result is often bankruptcy of businesses when the costs of production exceed revenue due to losing the competition for wages. The production of poverty occurs when living costs exceed wages in the low-wage markets. Migrant labor provides much of the semi-skilled and unskilled labor of construction and manufacturing, which results in vulnerability due to language issues, availability of work visas and the lack of worker protections and rights. When the race to the bottom fails to yield lower wages, corporations and states may turn to prison or slave labor. Prison and/or slave labor remains widely used in free market economies. For example, private prisons run by the Corrections Corporations of America contract with corporations to produce military uniforms for the US government.

FOCUS: SINGAPORE

Singapore developed its free market economy in the mid-20th century, and currently ranks as one of the least restrictive economies for business. It is also considered to be less corrupt and more open than other Asian economies developed around the same time. It has a high percentage of English speakers, high levels of FDI because of its location, a high percentage of skilled workers and low taxation. Foreign workers compose more than a quarter of the population; both men and women seek work in manufacturing, while men seek construction and shipping industry work, and women are often employed as low-wage domestic labor for middle-class workers (Kaur 2010). Singapore is one of the world's most expensive cities in which to live, but the competition from an influx of foreign workers suppresses wages and generates underemployment. The response from the Singaporean government has been to introduce limits on immigration and taxes on domestic worker employment. Changes in policy and the labor market have also triggered **circular migration**, or temporary migration in which workers seek work in a foreign country but return to their home country when demand for work is low. This pattern is common among male and female workers regardless of occupation, and the labor pool frequently contracts and expands to meet demand. Women are sexually harassed by their employers as a way to keep them vulnerable to firing, which includes pregnancy testing and routine screening for sexually transmitted diseases. Male workers in any profession are not required to undergo testing, however.

Planned economies

Planned economies operate under an economy-wide plan, in which investments and use of capital are used to determine what and how much of any particular good is produced in a given year. Planned economies take a variety of forms; they may be centrally organized by the state government, decentralized to other government entities, or some combination of the two. Another option is **participatory** planned economies, meaning they involve citizen or worker input on the plan. Many planned economies operate based on five-year plans, and democratic governments will include citizen referendums on economic decision making about the plan. The benefit of planned economies is that the state does not have to attract foreign investment for development, nor does the state have to wait years for profits to accumulate in the primary or secondary sectors in order to expand development. It also means that the state can seize assets, or dictate economic practices at any time, which is not beneficial or attractive to private investment and enterprises. The Soviet Union was one of the most widely recognized centrally planned economies, and other countries such as Chile, Venezuela, India and China have also used centralized planning.

Because planned economies are generally closed to trade and other forms of capital investment, waged labor is primarily characterized by internal migration, sometimes through coercion or other economic incentives to seek waged work. Commercialization in the agricultural sector does not always happen in a planned economy because capital is directed toward developing industry and manufacturing. As a result, wages are frequently suppressed due to the lack of competition in a labor market, and consumption is restricted to facilitate capital accumulation in the industrial sector. Basic needs are almost always met by the state – for example, food is sometimes provided to workers through **subsidies** (the cost is shared with the state through direct financial contributions). Healthcare and education are provided and coordinated by the state. Health and educational outcomes are consequently the same or better than those of free market economies. Housing in some cases is also supplied, such as worker dormitories in China's SEZs or centrally planned towns in the former Soviet republics. Labor is sometimes supplied through forced

Figure 5.2 Women's hospital in Havana, Cuba

labor, such as the Gulag system in Stalin's Soviet Union or North Korea's prison camps.

During several decades of the 20th century **Cuba** was part of the Soviet Union's planned economic system. Prior to forging political–economic ties to the USSR, Cuba had a single-resource economy (sugar cane), but the economy was characterized by high levels of wealth and investment, primarily from the United States, along with poverty and economic inequality prior to the Cuban Revolution in 1959. Poverty and unemployment triggered migration to the cities for work. Cuba exported sugar to many countries, and imported many of its agricultural inputs before the fall of the Soviet Union. Since the fall of the Soviet Union, Cuba is still dominated by state-run enterprises employing the majority of the population. In the early 1990s, Cuba reorganized its economy to focus on tourism, but retained its free education and healthcare systems along with food subsidies. The government sets most prices and rations goods to citizens; investment is restricted and requires approval by the government and poverty levels are now among the lowest in the developing world. Cuba is currently transitioning to a mixed economy with about a quarter of the population engaged in entrepreneurial activities. Cuba focused on racial and gender equality as part of the Revolution, and women enjoy advances in education, representation and healthcare, including generous maternity benefits and reproductive healthcare. Racial inequality, already entrenched in Cuban society, has increasingly eroded since the introduction of a dual currency system, which was meant to attract foreign capital. It has benefited the already privileged classes in Cuba, many of whom receive remittances from abroad (Blue 2007).

Mixed economies

A third path is through the development of a mixed economy. A mixed economy is one characterized by a "Third Way" of liberal economic policies accompanied by regulation that provides some check on the development of inequality and the negative externalities of capitalist processes. These include exposures to hazardous working conditions, environmental pollution and three crises of capitalism: inflation, stagnation and recession. Mixed economies assume that capitalism can provide a basic framework for an economy, and use some combination of private property rights and free enterprise, but employ fiscal, monetary or trade policies to mitigate the ways in which capitalism does not provide public goods. This strategy is often thought to employ the best of both free market capitalism and socialist central planning. Mixed economies have strong articulation with socialist democracies. For example, Norway is a contemporary example of a mixed economy. Market capitalism exists alongside state-owned enterprises, and the state provides a strong welfare system, and ensures high living standards. It should be noted, however, that Norway's ability to do this is premised on exporting fossil fuels, such as North Sea oil and gas.

Mixed economies are systems that incorporate a variety of government planning and intervention and free markets, including a variety of mechanisms for controlling the means of production – private property, public ownership, state-run enterprises, cooperatively owned and managed businesses. In many cases, mixed economies are market-driven with strong state oversight of business and state provision of public benefits, such as healthcare and education. Some mixed economies, however, feature state-run enterprises such as Bolivia's state-owned and -operated mining companies, or cooperatively owned business such as the Mondragon Corporation in Spain. One of the central philosophies of mixed economies is Keynesianism, which asserts that market fundamentalism often results in crises of capitalism (i.e., recession, inflation, unemployment), and in such cases state direction is required to salvage the economy and bring it back on track. The US financial crisis of 2008 signaled such a moment, in which banks and auto manufacturers needed "bailouts" of public money from the federal government to prevent failure and collapse of the domestic and international economies.

Polanyi (1944) speculated that such crises would happen in the 1920s. He argued that states and markets must always work together to stabilize the economy and ensure positive growth, although more planned and with slower growth than in free market economies. With the notable exception of states such as North Korea, the global economic trend continues toward mixed economies, wherein governments provide stabilizing forces on markets. Trends vary toward more planned or less planned, or facilitate private accumulation more than public goods, but economies generally are more stable when combined with democratic government regulation. This tends to produce better outcomes for everyone in terms of education and healthcare, but especially in terms of women's empowerment and acquisition of waged work, childcare and family planning.

The financial crisis of 2008 hit the economy of **Iceland** particularly hard, given that its economy was heavily reliant on the banking sector. It has since reorganized its economic activities, and like many Scandinavian countries is a prototypical mixed economy with high levels of government intervention combined with relatively free markets for export. It has a highly diversified economy as well, domestically self-sufficient in energy from hydroelectric and geothermal sources, reliant on exports of smelted aluminum and its extensive fisheries. According to the UN Human Development Index, Iceland is one of the most egalitarian societies, with high levels of participation by women in leadership, education and the workforce. While having some of the highest levels of gender equity in the world, women in Iceland still earn an estimated 14–18 percent less than their male colleagues. Iceland has a plan to eliminate the gender wage gap by 2022 by making employers prove they are paying women employees fairly. In March of 2017 on International Women's Day, Iceland's lawmakers introduced legislation that would require companies to prove they pay women and men in the same jobs equal wages.

Discourses of development

The development of an economy, regardless of its form, relies on the availability of labor. Different types of economies rely on divergent forms of labor, as suggested above, ranging from slave labor to highly regulated wage labor arrangements. The most common form of labor today is waged labor, which, if unregulated, can render workers highly vulnerable to exploitation. In what follows we outline the history and context of waged labor processes and explore the intersectional consequences of development that are embedded in the global capitalist economy.

Waged labor

Waged labor occurs in a situation where a worker enters into a contract with an employer to sell his or her labor time and power for some agreed upon amount of money. Waged labor relations often begin in a context of dispossession (see Chapter 4), where they are alienated from the means of production or subsistence activities. Van der Ploeg (2010) argues that states and capitalism work together to forcibly modernize subsistence economies toward consuming economies. The modernization process removes workers from primary economic activities as a form of subsistence or "self-provisioning" (e.g., making cheese to eat from milk produced on a farm) to manufacturing for wages, which then requires the worker to purchase the manufactured item with his or her wages. A waged laborer thus relinquishes control over the product to the employer, and his or her primary income comes from selling labor power. The employer accumulates surplus (or profit) when people are paid less than their labor is worth. Laborers' relative lack of power to negotiate wages (outside of the collective bargaining power of unions) and the cost fixity of other forms of production (land, inputs) results in the suppression of wages. This remains a frequent source of profit in a capitalist system.

Critics of waged labor arrangements argue that people are often compelled by economic necessity to engage in waged work, i.e., they are forced off land or out of agricultural work to seek waged work. They have no other feasible option, and, at this point, the choice to engage in waged labor is less a voluntary expression of free will, and more an exercise of economic power by private capital over the most vulnerable for the purposes of securing profit. This is referred to as the production of a **reserve army of labor**, or a class of unemployed people currently seeking work, who are often willing to work for wages that are lower than the current wage market. Take, for example, migrant workers from Haiti to the Dominican Republic: lacking legal status or rights as citizens, they work for less than minimum wage in the agricultural sector. Slavery and debt slavery are more extreme methods of securing profit by not paying wages at all, or by paying lower wages or trapping workers in debt. These forms of **exploitation** are more likely to be experienced in intersectional ways by women, migrants, people of color and those who are gender nonconforming. These groups are overrepresented in work that is exploitative, and are almost always paid less than white heterosexual **cisgender men** (or men whose gender identity corresponds with their birth sex), and experience a documented difficulty in negotiating equal pay for equal work even in privileged positions.

Globalization inaugurated a new era of production configurations by facilitating free trade agreements, such as NAFTA in 1992, which allowed for the movement of capital (i.e., FDI) to construct factories across national state borders, but not the free movement of labor. The point of such maneuvering is to search for the lowest wage in order to lower the costs of production. Where wages are low, worker protections (i.e., minimum wages, workplace environmental protections) are also poor or nonexistent, leading to exploitation or harm to workers. This movement of capital in search of low wages and fewer regulations results in what scholars call the **global assembly line** in which components of various products are produced in multiple locations and simply assembled in one place at the end stage of production. This is also referred to as a commodity chain or a production network. Women, especially young women, are often paid less than men for semi-skilled work, and are frequently enrolled in global assembly lines as what Wright (2006) calls "disposable workers."

Feminization of labor

Wright (2006, 2) argues that women are sought out as workers because they are assumed to be "dexterous, patient and attentive workers." Wright argues that they are also viewed by employers as "disposable" because they are easily fired and replaced when they are no longer physically or mentally able to carry on working under oppressive and dangerous conditions, or if they refuse to work in or organize a protest against their working conditions. A central theme and widely believed stereotype about women in the manufacturing sector is that they are "docile" and easy to control. This narrative has led to the feminization of manufacturing work since the late 20th century. Women workers' situatedness as uniquely vulnerable subjects, however, makes resistance or labor organizing difficult and dangerous. Ong (1991) calls this **flexible accumulation** or labor strategies reliant on women and migrants. These new arrangements shape the global assembly line and disrupt the development of a shared class consciousness and social movements for better working conditions. Ong (1991) argues that acts of resistance occur in everyday ways, such as working slowly, jamming equipment or taking long restroom breaks, rather than unionizing, which would drive wages up through the ability of collective bargaining. The logic that another willing, docile and dexterous worker will take her place, and who will also one day be disposed of, is one of the most dangerous practices of capitalism.

The **feminization of labor** is generally thought of as a general condition of work that devalues productive labor through deskilling, specialization and temporary work (Richer, 2012). Precarity of labor is often a result of these processes, which is experienced by both men and women, although women are generally more vulnerable to exploitation for reasons we discuss throughout the text. Flexible accumulation operates in the context of development focused on capitalist enterprises, which use a variety of control mechanisms to exploit labor in novel ways. Foreign direct investment allowed for the creation of free trade zones, spaces that are subject to little or no regulation of labor or the movement of capital. Women are constructed as desirable "docile" workers in these environments, and are enrolled in a variety of spatial forms of control and surveillance. Migrant women often live in dormitories segregated by sex, work on factory floors separated from each other, are not allowed to speak to one another, and their mobility is restricted both within the factory compound and in the wider community. Infractions are penalized by a loss of pay. According to Wright (2001), in Juarez, Mexico, the body and the social construction of a respectable girl is used to police behavior. When women do collectively organize to change working conditions, they are punished with sexualized violence and even murder. We examine this example of gender-based violence, free trade agreements and the feminization of labor in more detail below.

The combination of the globalization of assembly lines and the race to the bottom in wages means that some free market economies become stalled in the middle stages of the development trajectory. Stagnating wages mean high mass consumption is not possible, and the disposability of workers generates high turnover rates among workers, who are left with physical impairments that make it difficult for them to work again. The Rostowian development model depends on workers being paid what their labor is worth, and on the possibility for upward mobility in career paths, as well as safe working environments. The model breaks down when levels of inequality are high because people cannot afford to buy the products that they make because they are paid less than their labor is worth. A mix of development models to provide the most conducive environment for human welfare must not rely on exploiting labor in intersectional ways to export high-value products, while producing and consuming everyday needs domestically with a mix of free market mechanisms and government regulation.

Informal economies and housing

As our opening vignette suggests, many migrants, especially the men who are not always desired for factory work, may not find the waged work in manufacturing they were seeking. This can also happen when a surplus of labor is created through modernization in agriculture, thereby creating the reserve army of labor who are underemployed or unemployed. Many find work in the **informal economic sector**. Informal

Figure 5.3 Informal farmers' market in Uttarakhand, India

work is undertaken "off the record" in terms of regulation of the working environment by the state or city, or the taxation of wages. People are paid in ways that are referred to as "under the table" or "off the books." Informal work ranges from sex work and drug dealing to domestic work and construction. Work that is criminalized is almost always informal, and migrants who lack documentation typically work in the informal sector. Other kinds of work involve buying and selling food for a pop-up farmers' market at the edge of the city, for example.

The informal sector is important to the global economy because it is neither regulated nor taxed, so profit is more easily extracted from this kind of labor. People who work in the informal economy are vulnerable in place-specific ways, such as not being a citizen, being trafficked, or being in debt slavery. Even those working freely are subject to mandatory overtime without pay, lack of health insurance and sick leave, and may be laid off at any time without notice for any reason. Unions have been formed in response to this treatment, and ensure basic protections for workers, unless unionizing activity is banned, as it is in many places. The production of nonunionized labor is a key component of foreign policy, which promotes economic development, such as agricultural modernization, which forces people off the land and into precarious situations in cities. According to the International Labour Organization, it is estimated that half to three-quarters of all nonagricultural work in the least developed countries is in the informal sector. Informal work is also an important complement to the formal economy. The domestic or reproductive work performed by women earning no or low wages within a family suppresses wages for the wage-earning person in the household.

People who work in the informal sector also frequently live in **informal housing**, sometimes referred to as slums, shanty towns or barrios. Informal housing is characterized by a lack of basic services, such as running water, electricity or sanitation. The housing is often constructed as temporary when migrants arrive in a city, but because so many places experience a stalled development trajectory, temporary housing turns into a permanent informal community. Like some refugee camps that have persisted for decades, informal communities develop their own infrastructure, which they are at risk of losing if and when the city decides to clear the squatter community for capitalist economic development. Ananya Roy (2014) writes that a new development scheme in some poorer countries is to grant land title to squatters, so that they might sell their improved land to the next migrant and move up into the middle class. This rather utopian vision enrolls informal workers into the global capitalist economy through private property schemes and frequently marginalizes women, who have few rights or opportunities to own land.

Case studies: Dubai and Mexico

Above, we outlined the general trajectory of development in the global economy today and the gendered consequences of waged labor employment in the construction and manufacturing sectors of the economy. We focused specifically on how this has had an impact on migrants, particularly women, who migrate both internally and internationally for paid employment. According to Buckley (2012, 251), "migrants are often the first and most affected by economic crises; tending to be disproportionally represented both in highly cyclical sectors and in precarious forms of employment . . . and shouldering an 'unequal burden of risk' in contemporary processes of economic growth." In what follows we discuss the international migration of men to work on construction in the "**shock city**" of Dubai, and the internal migration of women who work in the maquiladora industries on the US–Mexico border.

Dubai, UAE

The economy of the United Arab Emirates has been rapidly growing since the 1990s, when it liberalized its economic policies, spurring economic development in the form of urban investment. The Gulf Wars injected foreign capital into the economy, and led to the creation of the Jabal Free Trade Zone. Dubai is often referred to as a "shock city," undergoing dramatic and regionally unprecedented growth from a population of just over one million in the 1980s to a glittering metropolis of nearly three million in 2017. Dubai's economy is centrally planned, but heavily supportive of free market policies and mechanisms. Tourism, finance, trade in petroleum and real estate are its top industries, placing it squarely in the global tertiary sector.

With its ambitious urban development in the form of luxury high-rise apartments, shopping complexes, human-made islands and hotels, Dubai faced a massive domestic shortage of construction workers, which was filled by migrants from South and Southeast Asia, including migrants from the Indian states of Kerala, Tamil Nadu and Andhra Pradesh. According to Buckley (2012), more than 700,000 workers lived in the city in 2006. The 2008 financial crisis gutted the sector, and many of these migrants returned home, only to return to Dubai when its economy improved. It is estimated that three-quarters of Dubai's population of approximately three million people are men, giving it the second-highest gender imbalance in the world (behind neighboring Qatar), and one of the largest migrant labor forces in the world (see Figure 5.4). More than two-thirds of its population are between the ages of 20 and 39, and an estimated 80–90 percent are noncitizens (Gottfried 2013).

Economic stagnation in India from the 1970s onward led to chronic underemployment facilitated by state neglect. Due to low levels of capitalization, anyone taking a loan for agricultural or other investment was forced into extremely high rates of interest. It took poorer, less skilled migrants many years to pay off debts, resulting in a form of domestic debt slavery in some states in India. Many men left India and traveled to the Gulf states as a result, thinking that high wages in either the formal or the informal sector would outweigh the high costs of visas and other illegal fees charged to migrants. Because so many workers are foreign, domestic unemployment rose rapidly within the UAE, resulting in state restrictions on foreign workers' ability to change employers, the length of work

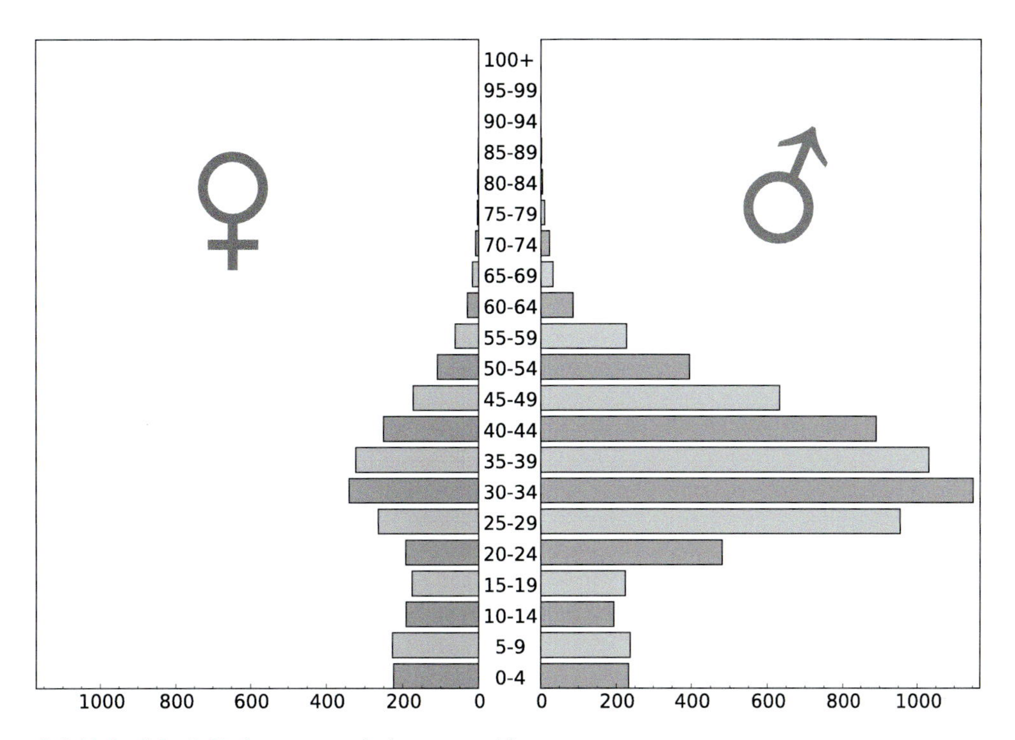

Figure 5.4 United Arab Emirates population pyramid

Source: by Central Intelligence Agency (CIA) [CC BY-SA 3.0 (https://creativecommons.org/licenses/by-sa/3.0)], from Wikimedia Commons

contracts and degradations to workplace conditions. And yet, Dubai continues to grow at a rapid pace. The demand for workers has not declined, but the promise of lucrative and safe employment for migrant men is no longer guaranteed, and many are returning to their homes.

Mexico

After decades of a failing import-substitution model, in 1994 Mexico signed the North American Free Trade Agreement (NAFTA) with the hope that it might revitalize its economy. This signaled an economic and political shift away from agriculture and toward manufacturing, which had gendered consequences. While NAFTA reordered relations between all three countries of the North American continent, what this meant for Mexico was the introduction of heavily subsidized agricultural imports, specifically maize (corn) – a cornerstone crop of the Mexican diet and economy. Farmers who could not sell their maize because it cost more to produce than the American corn were dispossessed of their means of production. This resulted in a massive international migration of men to work in agriculture and construction in the United States. Displaced and unemployed women were induced to internally migrate to newly established manufacturing facilities built by General Electric, Alcoa and DuPont on the Mexico side of the US–Mexico border.

As discussed above, women in the maquiladoras (and other export processing zones) are treated as disposable components of the production process, easily replaced when their bodies are no longer able to perform the "dexterous" work required. They will be disposed of through firing if they resist the real or imagined narrative about their identity as "docile" by attempting to rework or remake their conditions of work through union organizing. Women entering the maquiladora industry are also deemed "untrainable" for work, requiring higher levels of skill and education, thus rendering their labor time–space limited to their bodies' ability to endure 14-hour shifts, limited breaks, constant surveillance, pregnancy testing, threats of violence, sexual harassment and firing for slow work performance. While the average turnover rate is around 20–50 percent, rates of up to 100 percent each year are not unusual (Wright 2001). According to the corporations that employ her and the pervasive attitudes of the society around her, the maquiladora worker's value lies solely in her ability to provide value for capital through her undervalued labor, and in her ability to be disappeared and disposed of when her productive capacity begins to diminish through age or disability.

The debilitating and devaluing work that results in high turnover rates has also been linked to the murder of women and girls in Juarez, termed "femicide." Between 1993 and 2003, 800 bodies were found, most bearing signs of sexual violence and torture, with 3,000 still missing six years later (Sarria 2009). Framed as drug- and cartel-related violence, with blame placed on the victims, police and other officials suggest that the murders were the result of women and girls being in the wrong place at the wrong time. Given the pervasiveness of this notion, the police effectively dropped investigation of the cases in 2006 (Sarria 2009). Wright (2006) disputes this framing and argues that women and girls are murdered in Juarez simply for being women. Their presence in the city is a symbol of shifting social and cultural norms, in which women earn a living independently of men, which threatens the male supremacy of the cultural economy in Juarez. The notion that the victims' lives (and the lives of future victims) are not worth investigating as well as an inscription of the idea that women's lives are worth only what they can produce for capital is a narrative widely challenged by women in Juarez and beyond North America.

Future challenges

Marx (1867, 1977) warned that capitalism is prone to crises, and argued that the reserve army of labor is the most pressing challenge of capitalism. Surplus labor occurs due to the downward pressure on wages, which sparks a declining trend in levels of prosperity for workers, especially individuals marginalized because of their race, gender and class. The problem of capitalism identified by Keynes is instability and volatility in financial markets that negatively impact overall levels of wealth and prosperity, such as was seen in the 2008 financial crisis. For Polanyi (1944), the threat to stability arises from an increasing decline in the ability of social groups to organize and collectively advocate for worker protections and other forms of social welfare. From what we know about the way in which workers are intersectionally exploited in the global economy, any crisis of capitalism will have a destructive impact on the most vulnerable, triggering a positive feedback loop in economic disasters, migration and displacement. These groups are enrolled in the capitalist global economy through the mechanism of flexible accumulation. These three crises work together to render a large group of people economically marginalized during a general economic downswing, without the ability to form social solidarity or collectively organize.

Anthony Loewenstein (2015) describes how private interests (corporations, individuals) profit from the destruction of landscapes, livelihoods and communities. The US, UK and Australia are well-documented supporters of policies that allow private enterprise to profit from prison labor, refugees, prisoners of war, natural disasters and conflict. In nearly every case, the people from whom surplus value is extracted are intersectionally positioned as "others" in the global economy. For example, the US prison labor complex, which forces prisoners to work in manufacturing sectors while paying them below minimum wage, is estimated to be larger than the Soviet Gulag at its height. Other strategies include paying private contractors exorbitant amounts of public money to do the work of government agencies, such as disaster relief and refugee resettlement. A major trigger for migration is conflict, which forces vulnerable populations into places where they have few rights, and a desperate need for livelihoods and other basic needs. They are easily coerced into working for illegal levels of pay in difficult, dangerous situations. Conflict thus pairs with dispossession to generate a reserve army of labor, as well as capital in the form of resources that can be seized as a result of the conflict (see Chapter 9). This process brings the "model of for-profit government . . . into the ordinary and day-to-day functioning of the state – in effect, to privatize the government" (Klein 2007, 15).

In spite of, or perhaps in some cases because of, the production of desperate classes of labor, and the looming crises of capitalism (and other ecological catastrophes such as climate change), new forms of labor organization, economies and infrastructure are emerging. Manufacturing and construction alternatives in the form of **people-centered development**, which recognizes that economic development does not always result in positive outcomes for individuals and communities, are proliferating across various economies. Three essential elements of people-centered development include democratic accountability, gender equality and political sovereignty (OECD 2017). Alternative forms of employment, such as the Mondragon worker-owned cooperatives in Spain, offer more autonomy and control over working conditions and worker pay and avoid the problem of exploiting vulnerable labor (see Chapter 10). Worker sovereignty offers pathways for equal pay for equal work and an alternative pathway for profit. These alternatives offer the ideas of worker self-management, rather than supervised, hourly waged labor, and economic democracy as a philosophical alternative to capitalism. As indicated by the political sovereignty of people-centered development, as well as food sovereignty social movements, political solutions must be offered alongside economic ones, which is the subject of subsequent chapters.

Recommended reading

Behind the beautiful forevers, Katherine Boo; *The shock doctrine: the rise of disaster capitalism*, Naomi Klein; *Travels of a T-shirt in the global economy*, Pietra Rivoli; *Arrival city*, Doug Saunders.

Recommended viewing

Mardi Gras: made in China; *The power of community*; *Clothes to die for*; *Working women of the world.*

Questions for discussion

Why are mixed and some planned economies generally better at providing healthcare and education for women and children? Some people argue that factory work (however bad) is better than nothing; how could you counter that argument? Why do you think import substitution as an economic development strategy failed for most nations in the mid-20th century? Watch *Slumdog Millionaire* and read Roy's (2011) Slumdog cities; contrast the visions and narratives of development.

References

Blue, S.A. (2007). The erosion of racial equality in the context of Cuba's dual economy. *Latin American Politics and Society*, 49(3), 35–68.

Buckley, M. (2012). From Kerala to Dubai and back again: construction migrants and the global economic crisis. *Geoforum*, 43(2), 250–59.

Gidwani, V.K. (2008). *Capital, interrupted: Agrarian development and the politics of work in India*. Minneapolis, MN: University of Minnesota Press.

Gottfried, H. (2013). *Gender, work, and economy: unpacking the global economy*. New York, NY: John Wiley & Sons.

Kaur, A. (2010). Labour migration in Southeast Asia: migration policies, labour exploitation and regulation. *Journal of the Asia Pacific Economy*, 15(1), 6–19.

Klein, N. (2007). *The shock doctrine: the rise of disaster capitalism*. New York: Macmillan.

Lawson, V. (2014). *Making development geography*. London: Routledge.

Loewenstein, A. (2015). *Disaster capitalism: Making a killing out of catastrophe*. New York, NY: Verso Books.

Marx, K. (1867, 1977). *Capital, vol. 1*, trans. Ben Fowkes. New York: Vintage.

OECD (2017). *Shaping the 21st century: the contribution of development co-operation*. www.oecd.org/dac/2508761.pdf. Accessed 10/25/2017.

Ong, A. (1991). The gender and labor politics of post-modernity. *Annual Review of Anthropology*, 20(1), 279–309.

Polanyi, K. (1944). *The great transformation: the political and economic origins of our time*. Boston, MA: Beacon Press.

Richer, Z. (2012). "The feminization of labor." In *The Wiley-Blackwell encyclopedia of globalization*. https://doi.org/10.1002/9780470670590.wbeog201.

Rostow, W.W. (1960). *The stages of growth: a non-communist manifesto*. Cambridge: Cambridge University Press.

Roy, A. (2011). Slumdog cities: rethinking subaltern urbanism. *International Journal of Urban and Regional Research*, 35(2), 223–38.

Roy, A. (2014). Slum-free cities of the Asian century: postcolonial government and the project of inclusive growth. *Singapore Journal of Tropical Geography*, 35(1), 136–150.

Sarria, N. (2009). Femicides of Juárez: violence against women in Mexico. Council on Hemispheric Affairs. www.coha.org/femicides-of-juarez-violence-against-women-in-mexico/. Accessed 3/22/2015.

United States Department of Labor (USDOL) (2017). Geographic profile of employment and unemployment, 2017. www.bls.gov/opub/geographic-profile/. Accessed 2/13/2019.

van der Ploeg, J. (2010). The peasantries of the twenty-first century: the commoditisation debate revisited. *Journal of Peasant Studies*, 37(1), 1–30.

Wright, M.W. (2001). Feminine villains, masculine heroes, and the reproduction of Ciudad Juarez. *Social Text*, 19(4), 93–113.

Wright, M.W. (2006). *Disposable women and other myths of global capitalism*. New York, NY: Routledge.

6 Work, mobility and uneven development

Introduction

In 2010 Ellen Pao, an attorney who is now the CEO of Project Include, filed a gender discrimination lawsuit against her employer, Kleiner Perkins, citing a hostile workplace environment. Since then, repeated investigations into the culture of Silicon Valley have revealed a breathtaking pattern of sexual harassment and discrimination against women and people of color. This came to a very public climax in 2017, when an executive at Google was fired for circulating a memo in which he said that women were biologically inferior to men. In his estimation, this explained why women were not hired at the same rate as men in the information technology (IT) industry. On an epistemic level, very little seems to have changed since the 1960s when black women mathematicians working to put a man on the moon were forced to use racially segregated restrooms more than a mile away from their workplaces. While the women's bathrooms are no longer racially segregated and there are both men's and women's restrooms in all buildings today, the message from some workers and leaders in the technology industry is clear: women and people of color do not belong in this knowledge sector. Why does this pattern of racism and sexism against workers in both service and knowledge sectors persist?

From a global perspective, service and knowledge industries are relatively highly paid professions that are meant to facilitate high mass consumption. Alienation from the means of production and the commitment of time to waged labor mean people "buy" instead of "make." Inequalities, as we have shown in previous chapters, are frequently pre-existing; however, not everyone has the same opportunity to participate in waged employment in the manufacturing sector, and some people work to generate services and products in the tertiary sector. These include haircuts, food preparation, laundry – the unpaid work historically done by women within the household. As women enter the workforce in ever increasing numbers in nearly every economy, this work is frequently outsourced as part of the **new international division of labor** (NIDL), also known as the global assembly line, which has repercussions across space for intersectionally gendered subjects.

The central objectives of this chapter are to explain tertiary sector work and to demonstrate how workers are enrolled into service work in intersectional ways. Gender, race and nationality are key to shaping the workforce, as well as to explaining how and why certain kinds of work remain protected for the most privileged individuals in the economy. We discuss the ideas of formal and informal work, reproductive labor and gendered wage gaps. We argue that intersectional gender is an organizing principle in the development of a service sector: women, especially women of color, work in lower-waged "care" industries, while men work in "knowledge industries" for relatively high salaries. We discuss two case studies of Germany and

Indonesia to explain the relationship between migration and underdevelopment.

Service and knowledge economies

In keeping with the three-sector theory outlined in previous chapters, the third and final sector includes service work. The service (or tertiary) sector can include relatively low-wage work providing a service such as fast food, or relatively high-wage work in the "knowledge" industries such as IT. The relationship between development and the three-sector theory is that the more developed an economy becomes over time, the more workers will shift into higher-paid service sector work (relative to primary and some secondary sector work). Higher-income countries have a far greater proportion of the population engaged in service work than in primary economic activities such as extractive or agricultural work. The tertiary sector can include services to other businesses within the sector, such as logistics or marketing, or may involve the extension of an intangible consumable, such as live music performances. The focus in the tertiary sector is on the interaction between people, rather than an interaction between people and a natural resource or a commodity, as in the other sectors.

Service work is understood to be either the production or performance of something that is intangible. This could be a personal service, such as childcare, a public service, such as education, or a business service, such as logistics. While the product may be relatively immaterial (health), there are products that have a physical dimension: art, food or publications. Vast arrays of activities are encompassed by the term service work. They include activities that fall into four broad categories: 1) physical activities (e.g., meal preparation); 2) intellectual activities (e.g., teaching); 3) aesthetic activities (e.g., art, music); and 4) recreational activities (e.g., tourism). As mentioned, the social dimensions of service work distinguish it from manufacturing work, but it is also characterized by something referred to as "emotional" work or "intellectual" work, which articulates with race, class and gender. When we refer to "service," we are discussing typically low-waged work in food service, education, childcare and other services. When we refer to "knowledge" work, we are referring to typically high-salaried work in IT, computing, banking, coding, programming, engineering, medicine, etc.

Historically, service work (food preparation, cleaning, laundry, childcare, etc.) was performed by women in the **private sphere** of the household, or by slaves, the poor, migrants or racially marginalized women. This work was (and continues to be) unwaged and usually voluntary as part of the social and economic arrangement of heterosexual marriage. Also termed **reproductive work** by Marxists, it primarily functioned as a way of producing the next generation of laborers, for either the farming or the manufacturing economy. This changed in most developed economies during the mid- to late 20th century when social norms changed in response to economic pressures, and women began joining the paid labor force in nearly every economy, regardless of its stage of development. The entrance of women into the paid workforce was a result of a decline in wages overall, and the unique opportunity that sexism offered employers to suppress wages even further by employing women and paying them less than men who formerly occupied those jobs.

The concept of the **wage gap** is used to address the issue that women and women-identified people are compensated for their work in ways that are unequal to men. Equal pay for equal work is a rallying cry in protests advocating for the elimination of gendered wage gaps and for raising minimum wages. Low minimum wages are often identified as important areas of redress for gendered wage gaps because women disproportionately work in service sector jobs that pay poorly. Iceland is the first country to begin to dismantle systematic barriers to equal pay, and the World Economic Forum (2016) identifies Iceland as the most equal country according to its Global Gender Gap report. Explanations of the disparities often identify differences in work cultures, family obligations, gendered biases of managers, lack of maternity leave or lack of role models, without addressing how gender and race are organizing principles in waged labor markets. Those against attempts to raise minimum wages argue

Table 6.1 Top ten countries for gender equality, the *Global Gender Gap Report*, 2016

Global Gender Gap Index (out of 144 countries)	*Global rank*
Iceland	1
Finland	2
Norway	3
Sweden	4
Rwanda	5
Ireland	6
Philippines	7
Slovenia	8
New Zealand	9
Nicaragua	10

that people do not actually live off their wages and that it is instead supplementary income for teenagers or other dependants. The reality is that many minimum-wage workers across both the global north and south are single women, often mothers who require other forms of assistance (such as taxpayer-subsidized food programs) to survive on minimum wages, effectively subsidizing the profits of their employers.

The entrance of middle-class women into waged work also generated a demand for the services that women perform, such as food preparation and childcare. The frozen dinner is frequently invoked as the quintessential example of cultural change due to the entry of women into waged work. The time and energy required to cook for the family generated a demand for convenience foods. Childcare became a highly sought-after service as women juggled raising a family while developing a career. Maternity leave is widely understood to be central to retaining women in the workforce, as well as avoiding the problems of a declining population. As women enter waged work, they often delay childbearing, sometimes for as long as a generation – choosing to have children in their 30s rather than 20s. This results in a slowing of population growth, and a decline in the number of people willing and able to work. Every country in the world aside from Suriname, Papua New Guinea and the United States have some kind of policy that provides paid benefits to women when they are caring for a newborn (Rossin-Slater 2017). Women in manufacturing

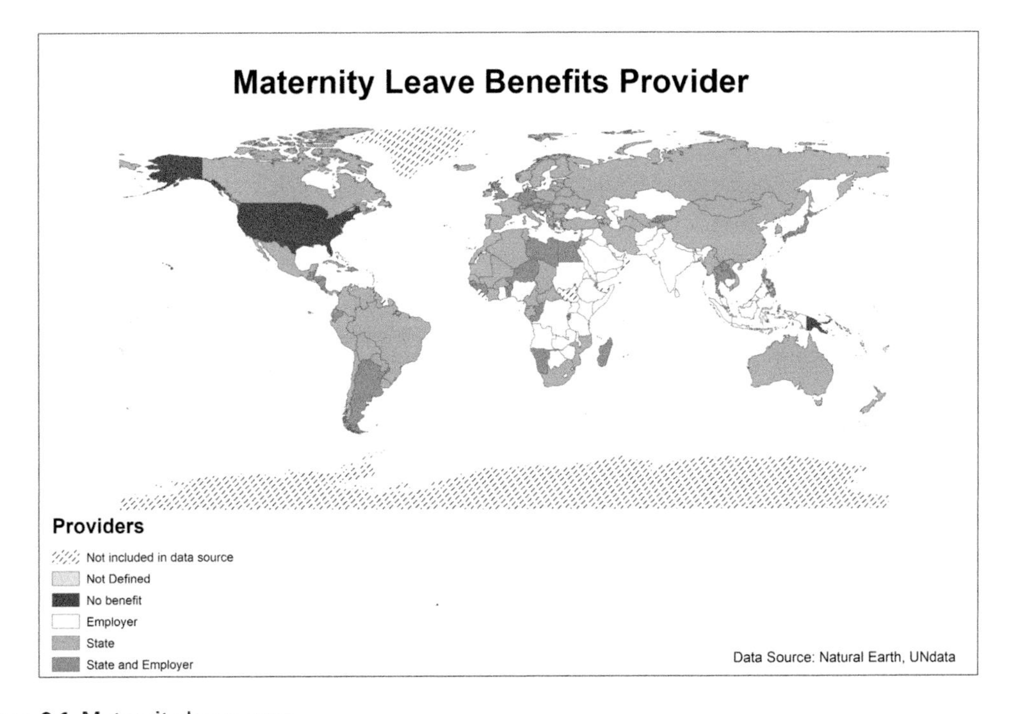

Figure 6.1 Maternity leave map

work are often subject to involuntary pregnancy testing, and women in primary economic activities, such as mining or agriculture, frequently bring their children with them.

There is a strong correlation with the availability in childcare services and women's participation in the workforce. Denmark, for example, has a program that includes up to a year of paid time off for both parents and four weeks of paid leave for the mother before the birth of the child. India provides paid time off for up to 26 weeks for women. While Germany has a very generous paid maternity leave (up to two years), it offers very little in the way of after-school care for children. This leaves women – who shoulder the most responsibility for childcare – to juggle informal childcare arrangements or holding part-time and temporary "mini-jobs." Recent studies indicated that more than a third of German women work part-time (Gottfried 2013). The absence or partiality of social safety nets in developed and developing countries alike forces women into strategies of "making do," often juggling unpaid childcare and other domestic work with paid work in what is known as the "third shift."

In the past few decades, labor markets have tended toward increasing informality and precarity, driven largely by the decline of collective bargaining, outsourcing, migration and the race to the bottom in wages. The most vulnerable workers are either working informally or are underemployed, taking wages that are far less than their labor is worth because it is either the only or the best option. The decline of corporate accountability to workers in the post-World War II period, as well as the development of a global assembly line, left more and more workers without guarantees. The massive offshoring and outsourcing surge in the 1980s led to the development of highly unregulated manufacturing work in spaces of dispossession, and the rise of service and knowledge sectors in zones of accumulation. Both processes employ the practice of intersectional exploitation of workers, but in different ways. It is important to remember that this is not an accident of capital, nor is poverty the inevitable outcome of low economic development; rather, it is part of a system to accumulate as much value from labor as possible and uses the process of globalization to normalize the degradation of working conditions and wages.

Much of **informal work** – or work that occurs "off the record" – encompasses a wide array of activities, including the drugs trade and prostitution, as well as domestic work or childcare. It is characterized by vulnerability and precarity, and is key to linking local economies to the global networks of capital. Informal work is difficult to measure because much of it is hidden. What is known, however, is that in some developing contexts, informal work is the norm, particularly for women, who must juggle childcare responsibilities while often having little or no formal education. Informal work is an important complement to formal work, as low- or no-wage work suppresses real wages, and often the "care" work done informally removes the obligation from the public sector to provide services, such as childcare. Informality is driven by a lack of state investment in workers and formal work, and leaves an already vulnerable group of people at the mercy of their employers. Informal workers, especially women and children, often work in the privacy of homes as domestic servants, cooks or nannies, and as informal employees; and they lack the protection of the state in illegal, dangerous or precarious work, such as sex work, mining or street vending. Efforts to formalize work have met resistance from both advocates and opponents of labor laws. Some groups claim informal work is one of few sources of financial security for very poor or migrant women, and formalizing it will make it impossible for them to enter those occupations. Others argue that development cannot proceed without access to rights and protections guaranteed by formal work.

As we have seen in every other economic sector, gender works in concert with race and nationality to organize labor and to differentiate between classes of workers. While the wages may be higher in the tertiary sector compared to other sectors, the subjective value added to the product or service through knowledge or emotion, and the intersectional vulnerability of workers, facilitates the uneven structure of the service and knowledge sectors. Women-identified people of color working in the informal service sector (e.g., migrant domestic workers) tend to be the most vulnerable and precarious, while white men working in the formal knowledge sector (e.g., information technology) tend to be the least vulnerable and enjoy the most benefits and rights. This kind of work is normalized, and considered aspirational, in spite of the way differently embodied people are systematically excluded and discriminated against and discouraged from working in those kinds of jobs.

Gender, service and knowledge work

McDowell and Court (1994) suggest that all forms of embodied service work, regardless of profession, add up to a kind of sex work, i.e., that the kind of **emotion-work** (labor dedicated to produce feelings) that women perform in many kinds of service work (childcare, nursing, domestic work) is attributable to the social construction of gender based on ideas of a binary **sex–gender system**. This concept was coined by Gayle Rubin (1975) to explain how gender roles emerged from the role of biological differences in the social reproduction of societies. Emotion-work converts the display of empathy or care into a **use value** (the value obtained through use, rather than buying and selling), which is exchanged for wages. While a manufacturing laborer converts energy into a product, the emotional laborer creates value from the transmutation of her feelings – by producing a feeling in the recipient (a child, a client, a patient). "The service worker, like her counterpart in the factory, decides how much emotional labor to give (work effort) and how much to withhold. Unlike her counterpart, she produces a sentiment rather than an object" (Gottfried 2013, 81). Thus emotional labor creates a commodity – something that is bought and sold – out of feeling, which is also economically devalued and dovetails into the way in which race and gender are incorporated into capitalist wage systems that produce inequality. Women's relative lack of access to power in the global economy makes them vulnerable to the way in which capitalism seeks to pay laborers less than their labor is worth.

Just as gender is a social construct, it is also used to represent workers in certain intersectional ways. For example, women are thought to be better suited to service work, while men are thought to make good knowledge workers. According to Gottfried (2013, 76), "eliciting emotional responses" from customers and fellow workers is a key aspect of service sector work. Much like women are deemed to be docile workers in manufacturing, women are seen to have better "soft skills" that compose the emotional labor skill set. The body is managed in a manufacturing setting, while the person and affect are managed and disciplined in the service sector to condition certain outcomes. Grosz (1994, 238) argues,

> power is inscribed on and by bodies through modes of social supervision and discipline as well as self-regulation. The bodies and behaviors of individuals are targets for the deployment of power, and they are also the means by which power functions and proliferates.

Behavior is controlled by making the body into "a particular *kind* of body . . . so as to make [it] amenable to the prevailing exigencies of power" (Grosz 1994, 239). The body is thus positioned within social narratives in space, and behavior is controlled through discourses of what is "appropriate" or "normal." In terms of gendered work in the tertiary sector, bodies are expected to conform to the norms of service and knowledge work, and their performances are compensated and punished in accordance with gendered expectations.

Perhaps the service most dependent on the embodiment of a particular gender identity is sex work. One of the most criminalized and least regulated service sector activities, sex work is widespread as both legal and criminal, voluntary and involuntary, sometimes within the same country. For example, sex work is legal in a limited way in some US states, but is part of black market economies everywhere in the United States, regardless of the law. This parallels the way that sex trafficking is often a product of conflict and civil breakdown, or where the state lacks the authority to enforce the law. Legal sex work, typically in the socialist democracies of Europe, indicates that worker protections save lives and improve the lives of sex workers, largely through its visibility and spatiality in buildings rather than on the street. One of the most overlooked forms of sex work is that for the disabled. For example, in the US state of Nevada, where prostitution is legal, sex workers are specifically trained to facilitate the sexual experiences of men and women confined to wheelchairs, or living with a wide range of developmental or physical disabilities, or chronic health issues.

Men also perform emotional work in the service sector, but they are more likely to be employed in "knowledge" industries as managers, doctors or executives, where aggression or emotional distancing is valued. Because of the binary nature of the sex–gender system, this normalizes and valorizes certain kinds of behavior for men, such as displays of anger, and pathologizes and devalues them for women. It also makes the emotional labor that men perform less visible, or less recognized as such, and makes it difficult for more empathetic or nurturing men to be taken seriously in such jobs, thus perpetuating a certain kind of hegemonic masculine work culture, particularly in medicine or technology. As long as this persists, it will be difficult for women to advance to higher levels of management in certain professions. The differential between expectation and reward for gendered emotional labor perpetuates gendered inequities and limits women's access to certain kinds of well-compensated work, thus perpetuating the gendered wage gap. This gendered work culture is also known for producing limits in upward mobility for women, also known as the **glass ceiling**.

For example, nearly 75 percent of computer programmers, one of the highest-paying jobs, on average, in the United States, are men. Women who work in knowledge industries are overrepresented in data entry jobs as well as concentrated in healthcare and social work. The STEM fields (science, technology, engineering and math) in the United States, and in most other countries where STEM is a large and important sector, remain dominated by men in the highest-paid positions and most masculinist work cultures. In the United States, as well, the IT sector is heavily reliant on importing labor from other countries, with nearly three-quarters of the workforce foreign-born. Countries such as Iran, where STEM education and occupations are not so specifically gendered, have higher rates of participation by women. Research in the UK indicates that these professions are also relatively closed to young people and others without access to a variety of social capital (education, wealth, locational advantages) (Allen and Hollingworth 2013). The loss of corporate accountability to workers has meant that these jobs are increasingly characterized by work contracts that are temporary and contingent on other market forces, which depresses wages in very high-value, high-wage industries.

Gender, work and migration

Borders and boundaries and the state system of territorial control work to differentially position subjects vis-à-vis borders and economic development. Citizens have a right to certain kinds of protections and privileges, while noncitizens do not. This is further divided on the basis of gender and race. Migrants of color and women are often positioned weakest relative to tenure and wages, and experience higher levels of precarity and vulnerability to exploitation. Huang and Yeoh (2003) demonstrate with their case of women workers in service industries in Singapore that, like all racism and sexism, this discrimination is capricious and arbitrary. Privileged Singaporean women who are citizens join knowledge economies as part of the middle class, and hire domestic workers from abroad to preform reproductive labor of the household. They may be technically the same race, ethnicity or religion, but frequently discriminated against because of their citizenship status (Huang and Yeoh 2016). Thus, Singapore has an extensive foreign labor market (at all skill levels), but inequity is produced through the expectations that foreign workers are temporary rather than remaining in Singapore long-term.

One of the largest gendered contemporary migrations is the flow of domestic workers from South and Southeast Asia (the Philippines, Indonesia, Sri Lanka, Thailand and Myanmar) to the rapidly developing economies of Hong Kong, Singapore and Malaysia. The trend of gendered employment in Singapore is for men and women to migrate from the aforementioned areas to work in manufacturing and service work (respectively), as both Singaporean men and women enter the burgeoning tertiary sector (information, finance). In Singapore, migrant women play key roles in replacing the reproductive work that local women used to perform in the home. Huang and Yeoh (2003, 23) write that this rests on the "idea that reproductive labor is not only commodifiable, but that it can be transnationalized, that is bought and sold across national borders." Rather than firms going in search of labor in other countries, employers seek to import labor to work in the home. They go on to note that while the opportunity to work in another country may seem like a chance to escape underdevelopment in one's home country, the "work merely transforms the nature of women's subordination rather than provides the means for alleviating it" (Huang and Yeoh 2003, 24). This works through established ideas about gender and nationality, and reinscribes them through the mechanisms of rights and borders. For example, in Singapore, a migrant woman may employ another migrant with a similar ethnic and religious background as a domestic worker (e.g., an Indonesian Muslim maid working for a Malay Muslim employer). The rights associated with residency versus work permits or lack of documentation mean that differential treatment arises from perceptions of belonging and class positions, leading to disparate forms of vulnerability that are enforced by the legal categories of belonging (permanent, temporary, illegal).

Immigration is often seen as a solution to a shortage of workers, and national states open and close borders in response to periods of unemployment. Immigrants often migrate to higher-income countries to seek work in the service sector, and knowledge workers are highly sought after. Migration results from uneven development of the global economy in which dependency theory explains that development in one place happens as a result of **underdevelopment** in another place. Underdevelopment includes the lack of investment in home communities (e.g., Indian migrants going to UAE) or negative impacts of trade liberalization (e.g., Mexican women in maquiladoras), while underemployment of rural populations leads to an out-migration of people in search of work. Uneven development is also a driving force in the new international division of labor in which economic sectors are segmented by costs of living and wage markets. Migrants, as we have discussed in earlier chapters, are frequently seen as either temporary workers, who will return to their home countries in a pattern of circular migration, or second-class citizens who are not afforded the same rights as citizens, even when they are on a path to citizenship. Egregious examples of human rights abuses against migrant workers abound in nearly every economy, and include forced labor, sexual harassment, denial of workplace safety standards (such as the use of a restroom) and the threat of deportation.

Case studies: Germany and Indonesia

In what follows we discuss two case studies that demonstrate the intersectional impacts of international and internal migration. Migrants in these cases left situations of underdevelopment in primary sectors in order to find work in more highly developed economies in the secondary and tertiary sectors. In the case of Germany, we examine the gendered impacts of the migration from Turkey, over a generation, as young women assimilate and reject the patriarchal norms of their immigrant families. In the second case, we examine the gendered impacts of differential access to rights when women migrate from other parts of Southeast Asia into the service sector of Indonesia. We show how gendered patterns of migration are central to certain regimes of work and development of new economies, and may have cultural and political implications for generations.

Germany

A case study from Germany indicates how gendered migration patterns result in both intergenerational migration and intra-household conflict among migrants over time. During the postwar reconstruction era in West Germany, a booming labor market due to a robust economy, low birth rate, an aging population and low unemployment led to a demand for migrant workers during the 1950s and 1960s. On October 30, 1961 the West German government signed a treaty designed to facilitate Turkish migration, largely due to a long history of positive political and economic relations between the two empires and countries since the 19th century. Turkey had a large surplus of relatively uneducated, poor rural men (a reserve army of labor), which was not available in any other European countries at the time. The West German government sought healthy unmarried men for their guest worker program for two-year stints to work at poorly paid jobs in factories. They hoped that these workers would fuel the "German miracle," a postwar development scheme aimed at rebuilding the economy. The guest worker programs, focused on men, were designed to prevent permanent immigration and to facilitate circular migration. The program was eventually changed to allow families to stay as workers and employers advocated for longer contracts and the right to permanent residence.

Turks now comprise the largest minority ethnic group in Germany, most of whom are Muslim, leading to the construction of cultural infrastructure such as mosques, Muslim community centers and schools. Similar to other immigrant communities, intergenerational conflicts arose in this community amid the older generation's desire to preserve sociocultural practices and the younger generation's interest in assimilating within German society. Patricia Ehrkamp's (2013) work reveals how migration has an impact on the social construction of gender. For example, daughters of immigrants selectively conform to and resist the gendered norms expected from their parents, particularly around issues of veiling and controlling their own mobility. Many young women express a keen desire to rework gender roles within their households and in their adopted country. Germany, due to its experience with Turkish migration, has worked to encourage continued migration, claiming it will take up to one million Syrian refugees. It also seeks to fill hundreds of thousands of highly skilled jobs in science, technology and engineering, aiming to employ up to one million people in the sector by 2020. It has subsequently eased rules for entry, and is one of the easiest states to enter in the European Union. While it has since revised its immigration policies to be more family friendly, it will have to revise its childcare provision in order to enable more young Turkish-German women to work full-time if they wish. The German government is further tasked with confronting a nationalist backlash in the wake of several terrorist attacks in various European cities.

Indonesia

Rachel Silvey's (2007) research indicates the importance of women's informal work to the survival of rural households in Indonesia for economic development. The financial crisis of the late 1990s in Indonesia led to the significant shift of a large population of workers from urban centers to rural areas. The Indonesian state encouraged this migration in order to quell unrest and to absorb the shocks of recession. Before the crisis, Indonesia was investing heavily in industrial processing zones, employing young women from rural areas in a common employment pattern across countries. According to Silvey, 63 percent of the workers in export processing zones were women who provided remittances, as well as food, to their home villages. Women became so important to Indonesia's economic development plan that they were identified as "heroes of national development." Migration was key to the survival of rural households because they lacked the resources to meet basic needs.

The financial crisis resulted in the unemployment of these young women, which had an impact on nearly every household. The return of women workers to their villages of origin was marked by intergeneration and intra-household challenges and contestation. The social construction of gender in rural Indonesia typically marginalizes young women from spaces of decision making, imposes restrictions on mobility and prioritizes men's health and nutrition. Women who had migrated to work in cities observed a cultural shift and participated in new forms of autonomy regarding money, work and mobility. When they, and frequently their newly unemployed male relatives, return, they find a gendered system of social relations that more often than not relegates them to unwaged reproductive work in the household. Silvey (2007, 42) writes, "Relative to their male counterparts, young women tended to carry heavier workloads and have less access to household resources, including cash and food." The idea that the rural households would provide social safety nets for the unemployed ignores how "returning . . . has made their lives more difficult than they otherwise would have been if household hierarchies were organized more in their favor" (Silvey 2007, 42). The Indonesian state's unwillingness to provide for unemployed workers and ensure corporate responsibilities to workers relies on the gendered devaluation of women and their work, and the gendered social systems they support require women to absorb financial shock by going without.

Silvey's (2007) research on Indonesian domestic workers in Saudi Arabia examines the ways in which the Indonesian state encouraged women's domestic labor abroad by identifying them as "Heroines of National Development." Many Indonesian domestic workers in Saudi Arabia endured significant hardships (including abuse from employers). Despite public knowledge about the abuse workers experienced, women continued to participate in this work. Most Indonesian domestic workers secured their travel to Saudi Arabia through a Hajj visa. Hajj, the pilgrimage to Mecca, is one of the five pillars of Islam. It is a coveted and meaningful part of Islamic praxis. Therefore, many Indonesian Muslim women sought to improve their social status upon returning home, partly because of remittances sent home, but more so because they enjoyed the honorific title of Hajjah (a woman who has gone on Hajj). Completing Hajj provided these women with more respect and social standing, which for them was more meaningful than the income they received as domestic servants.

Future challenges

In Chapter 4 we discussed how capital moves across borders to seek the lowest-waged workers. In this chapter we discussed how people move across borders to seek work, and how that leaves them vulnerable to capitalist exploitation in a variety of intersectional ways. Geopolitical conflict and economic recession also contribute to massive waves of migrants, which strain resources and create humanitarian crises. The world's borders are becoming ever more constrained due to crises in the legitimacy and effectiveness of the state system. For example, Europe, Canada and the United States began restricting migration across borders because of concerns about migrants who were perceived to be threats to national security. The reality is that domestic terrorism is largely carried out by citizens or people on student or permanent visas, and denying immigrants and migrant workers entry is a grave economic mistake. Refugees and guest workers rarely carry out acts of violence, and the vulnerable – those fleeing violence or instability – are the most in need of social welfare provisions and work permits. Denial of legal residency and work permits robs people of dignity as well as opportunities to assimilate and gain autonomy, and inevitably sows the seeds of discontent among the next generation.

An intersectional gender analysis of these processes indicates that women are disproportionately affected by limits placed on migration for economic, asylum and family reunification reasons. Women from **spaces of dispossession** are the least likely to commit acts of terrorism and the most likely to contribute to the capitalist accumulation of surplus through their suppressed wages in **zones of accumulation**. They also supply much of the unpaid labor to reproduce a population of workers. We also find that intersectionally embodied women and men are the most vulnerable in a system which ties rights (citizenship) to territories of birth and nationality. Migrants are uniquely vulnerable to exploitation, and their intersectional embodiment (i.e., presumption of illegal entry on the basis of race) is frequently used to justify their oppression and low wages. The trend of outsourcing work from the 1990s onward is reversing course now, as recessions constrict economic development in the zones of accumulation, and new forms of development enable new opportunities for emerging economies. For example, India's and China's burgeoning domestic information and service sectors, in which second- and third-generation internal migrants (both men and women) are working, are contributing to the development of a new middle class. Tertiary sector work is changing the work culture in both negative and positive ways. "Flexibility" – which we would use to classify temporary and precarious work – characterizes new working arrangements, allowing more time for work at home or working hours that are more compatible with caring for families. This comes as the workforce feminizes, as men work more at home, and women gain more power over the conditions of their work. Tertiary sector work is also precarious, and the more vulnerable in intersectional ways (migrant, minority) the worker is, the more he or she may be made vulnerable to exploitative employers.

While the emergence of new geographies of work may seem beneficial to some, such as the proponents of the "creative economy," others argue that this is just window dressing on the rollback of state and corporate responsibilities to workers, as the wealthy continue to gain profits from paying workers less and less. This mirrors the reliance on informal work in the least developed economies, as well as the persistence of gendered work cultures in the most developed countries. These patterns indicate that, regardless of context, gender, in concert with race and ethnicity, is used to marginalize the most vulnerable and extract value from their labor by paying them less than their labor is worth through intersectional hierarchies of labor value. The service and informal work that women, across cultures, continue to do at much higher rates than men (although men also do informal work) indicates that the lack of social safety nets perpetuates precarity and vulnerability.

Recommended reading

Nickeled and dimed, Barbara Ehrenreich; *Map of the world*, Randa Jahar; *Cities of salt*, Abdul Rahman Munif; *End of loyalty*, Rick Wartzman; *Reset: my fight for inclusion and lasting change*, Ellen Pao; *The McDonaldization of society*, George Ritzer

Recommended viewing

Hidden figures; *Beyond borders*; *Intouchables*; *A day without a Mexican*

Questions for discussion

Where do you see informal work taking place around you? What role can you see it playing in the economic development of the United States? Where do you stand on the formalization of work arguments? Read Collier and Betts's (2017) article on refugees. Why do they argue that limiting migration is a catastrophic error? How does this relate to economic development? Watch the 2007 documentary *Beyond borders*, which asserts that migration is a human right and necessary for the maintenance of populations in developed countries. Craft arguments for and against immigration on the basis of economic development, demographic transition and human rights. Be sure to include how migrants would be protected from exploitation. Identify three services provided through the work of a migrant that are central to your daily life. Go without them or try to do them yourself.

References

Allen, K., & Hollingworth, S. (2013). "Sticky subjects" or "cosmopolitan creatives"? Social class, place and urban young people's aspirations for work in the knowledge economy. *Urban Studies*, 50(3), 499–517.

Collier, P., & Betts, A. (2017). Why denying refugees the right to work is a catastrophic error. *Guardian*, March 22. www.theguardian.com/world/2017/mar/22/why-denying-refugees-the-right-to-work-is-a-catastrophic-error. Accessed 2019/3/25.

Ehrkamp, P. (2013). "I've had it with them!" Younger migrant women's spatial practices of conformity and resistance. *Gender, Place & Culture*, 20(1), 19–36.

Gottfried, H. (2013). *Gender, work, and economy: unpacking the global economy*. Cambridge: John Wiley & Sons.

Grosz, E. (1994). *Volatile bodies: toward a corporeal feminism*. Bloomington, IN: Indiana University Press.

Huang, S., & Yeoh, B.S. (2003). The difference gender makes: state policy and contract migrant workers in Singapore. *Asian and Pacific Migration Journal*, 12(1–2), 75–97.

Huang, S., & Yeoh, B.S. (2016). Maids and ma'ams in Singapore: constructing gender and nationality in the transnationalization of paid domestic work. *Geography Research Forum*, 18, 22–48.

McDowell, L., & Court, G. (1994). Performing work: bodily representations in merchant banks. *Environment and Planning D: Society and Space*, 12(6), 727–50.

Rossin-Slater, M. (2017). *Maternity and family leave policy* (No. w23069). National Bureau of Economic Research. Cambridge, MA: Harvard University Press.

Rubin, G. (1975). "The traffic in women: notes on the 'political economy of sex.'" In R.R. Reiter (Ed.) *Toward an anthropology of women*, New York, NY: Monthly Review Press (pp. 235–236).

Silvey, R. (2007). Unequal borders: Indonesian transnational migrants at immigration control. *Geopolitics*, 12(2), 265–79.

World Economic Forum (2016). *Global gender gap report.* http://reports.weforum.org/global-gender-gap-report-2016/. Accessed 10/26/2017.

Part III

Moments in development

7 Health and population

Introduction

The relationship between health and population is explored in this chapter through the lens of environmental and workplace health, reproductive health and the politics of population control. Poor or declining health is by definition an embodied experience and therefore we explore this through the lens of gender, race, class and sexuality.

Disparities in health and access to healthcare for both men and women are place-specific and related to differences in social status based on income, ethnicity, race, disability and sexual orientation (WHO 2017). Individuals who are marginalized because of their race, ethnicity, socioeconomic class, sexual orientation or political affiliation are often forced to live in unhealthy environments, such as areas with contaminated water, soil or air. Endemic and entrenched poverty, racism, classism and other forms of sociopolitical marginalization lead to various forms of ill health, and marginalized populations have historically been targets of governmental population control programs. Global inequalities as discussed throughout this book are further highlighted when we examine health outcomes, healthcare provisions and access to both preventive and prescriptive healthcare. The following section further explores the relationship between intersectional gendered experiences of work and environmental conditions, and their impact on individual and collective health outcomes.

Gender, work and environments

Economic development programs have led to increased economic globalization and the growth of factory labor in locations with an abundant low-wage workforce. These locations are also desirable to corporations because there is little to no oversight, regulation or unionization (Sparke 2013). Living and working in toxic or unsafe environments or areas without clean water and air has a significantly negative impact on both laborers' and the public's health. Globalization has included a relatively rapid migration of individuals from rural to urban areas, which has been challenging for many cities because their existing infrastructure cannot meet the needs of the population, particularly access to safe drinking water. Systemic neglect of migrants, corruption and lack of authority (or existing capital) to leverage municipal funds for infrastructure are further culprits of inadequate public health programs. Lack of access to safe water has an accumulated negative effect on individual health, which varies significantly by place and gender. In many places without running water in the household, women and children are tasked with collecting water for their families. When water is scarce or must be found at a greater distance from one's house (in either urban or rural areas), obtaining water adds additional time-to-labor burdens for women (Davidson and Stratford 2007, O'Reilly et al. 2009, Reed and Christie 2009). Hauling enough water for a household can be very time-consuming, preventing women from completing other chores in or outside the household (Buor 2004). Contaminated water has more negative effects on women as compared to men, especially in places where women are tasked with responsibilities that put them in more regular contact with waterborne toxins, such as washing and cooking (Sultana 2011). Additionally, water provisioning has an emotional impact on women (and sometimes men) because they cannot always be certain of the quality of water available to them and their families (Sultana 2011). The cost of fresh water can also exceed the cost of food or other necessities in some urban areas, which makes accessing water overwhelmingly difficult for the economically disadvantaged.

Water and sanitation

Ensuring access to safe water is inextricably linked to efficient and effective sanitation practices. Water, wastewater disposal and sanitation have become incorporated into development through efforts to improve infrastructure and health/healthcare programs. Women's sanitation needs differ from men's due to menstruation, pregnancy and childbirth. Therefore, when access to toilets is limited, women are at a much greater risk of contracting illnesses and infections than men. In other cases, when bathrooms are not available within one's home, women risk harassment or assault, particularly if they are seeking to use a bathroom during the night. Additionally, waiting extended periods to urinate (or limiting fluids in an effort to avoid urination) can lead to health problems such as urinary tract infections and kidney stones. Sanitation projects have become an integral part of economic development programs; however, they tend to focus on technological fixes rather than providing sanitation that will work well for all members of a given community. Thus, sanitation development projects need to incorporate and understand local attitudes, including gender roles and relations. For example, in some places "gendered ideas of privacy and self-respect [have been] capitalized on to change behavior" (O'Reilly and Louis 2014, 48). Ensuring women's privacy in public toilets has been an effective method for initiating sanitation improvement projects. Incorporating gendered sanitation and hygiene with existing sociocultural practices has also proved to be an effective method for safeguarding and improving public health. Therefore, education programs that address sanitation and hygiene have been found to be effective at improving the overall health of a community (Halvorson 2004). In some cases, education can help to improve sanitation practices within communities, when infrastructural improvements are slow or nonexistent. Additionally, technology may help to improve access to potable water or improved sanitation; however, without community understanding, input and knowledge, these new technologies can create more problems than they solve.

Lack of potable water and waterborne illnesses remain major factors in maternal health and the health of infants and small children. Diarrhea and other waterborne illnesses account for a significant portion of childhood morbidity, particularly in areas without regular access to safe and clean water. In locations without access to drinkable water, capitalist interventions have taken the shape of paid water services, effectively turning water into a commodity. Paid water services in many cases offer a solution, but these services highlight significant problems with one's ability (or lack thereof) to purchase water. For example, paid water services in Ghana were used to supplement water in urban areas with limited access to safe water. These fee-based water options provided safe drinking water to individuals, while simultaneously directing public pressure away from the government to improve the infrastructure that would ensure safe water (by way of public utilities) for all citizens (Stoler et al. 2012). Water quality (similar to air quality) remains a major public health concern in various geographic locations. The growing prevalence of market-based solutions, such as individual air and water filtration systems, attends to the needs of individuals in a particular household who can afford these products, while ignoring or placing less pressure on governments to improve conditions publicly for all citizens, irrespective of their ability to pay for clean water or air (Williams and Mawdsley 2006). Capitalist and market-driven systems to ensure public health benefit some, but not all, and further direct attention away from governments and towards the privatization of public health. Thus, it is imperative for public health programs to view all members of "the public" equally rather than divided by social categories such as gender, race, sexuality, class and dis/ability. Privatizing public health exacerbates health disparities by way of economic inequalities and jeopardizes the health of the environment.

Labor

In addition to water and air quality, different forms of physical labor occur in unhealthy environments. Laborers in these unhealthy environments work for low wages in unregulated factories. Many of these individuals also experience social marginalization or exploitation associated with a combination of their social class, status, race, gender, sexuality or dis/ability. Other forms of exploitative work rely on transient or migratory labor populations without the citizenship rights or ability to advocate for improved work conditions. Migratory labor tends to follow certain gendered patterns. For example, women dominate the migratory labor sector for domestic work (i.e., housekeeping and childcare), and many of these workers have experienced physical and sexual abuse at the hands of their employers (Silvey 2004, Pratt 2012). The lack of migratory labor regulation (particularly domestic labor) and the inability of migrants to advocate on their own behalf keep migrants locked in cycles of risk and abuse, with both short- and long-term effects on their physical and mental health and overall well-being. The negative health effects in industries such as mining, road construction and heavy manufacturing are predominately experienced by men, as they are the primary laborers in these industries. In many countries without a strong economy, surplus labor has become an attractive prospect for global companies to increase their profits by hiring low-wage laborers.

Many migrants seek employment as a primary driver for moving from one place to another. Some forms of migratory labor that are dominated by men include physical risks and fatalities. "For example, in the U.S., the three occupational groups with the highest rates of occupational fatalities (transportation, construction, and agriculture) . . . [have the] highest proportion of immigrant workers" (Schenker 2010, 3). Men are the majority of the workers in these professions and therefore most workplace injuries and fatalities are experienced by men, with only 8 percent of fatalities occurring among women (Schenker 2010). Men in migratory labor professions such as road construction often experience long-term health effects such as respiratory illnesses and infections (Sabhlok, Cheung and Mishra 2015).

In other places, both men and women are at risk of workplace injuries and fatalities, particularly when workplace environments such as factories are not safe spaces or are improperly built. For example, workplace fatalities occur in various industries, from clothing manufacturing to information technology. The pressure placed on factories by corporate owners to lower costs for labor, materials and factory maintenance has led to poor construction of factories and abusive treatment of employees. For example, poorly constructed and maintained garment factories in Bangladesh led to the collapse of a Rana Plaza factory in 2013 which killed 1,134 workers and injured approximately 2,500 (see Figure 7.1). Other forms of work-related problems include the abuse of alcohol and other drugs by employees who work long shifts or engage in labor that strains or injures their bodies. For example, men who participate in various forms of hard labor may use alcohol or other drugs to mitigate the bodily pain they experience from this kind of work. The use or abuse of alcohol and drugs may have added negative health effects on women and children within the household who experience physical abuse at the hands of inebriated male heads of households (Cleaver 2002). Therefore, efforts to economically develop a place have had seriously negative health consequences for the individual bodies of men and women whose labor produces the infrastructure necessary for global capitalism to flourish, ensuring the continued wealth of the few through the economic and health exploitation of the impoverished many. This exploitation includes systemic short- and long-term health problems for laborers. Healthy work and living environments are also an important component of reproductive health.

Figure 7.1 Rana Plaza building collapse, April 24, 2013, Dhaka, Bangladesh

Source: by rijans (Flickr: Dhaka Savar Building Collapse) [CC BY-SA 2.0 (https://creativecommons.org/licenses/by-sa/2.0)], via Wikimedia Commons

Reproductive health

Living and working in a healthy environment with access to affordable and quality healthcare is inextricably linked to improving women's reproductive health. Reproductive and maternal health practitioners also seek to improve women's health by reducing the number of births per woman and amount of time between births. Women who have several children without much time between births are at greater risk of illness or death due to the strain of pregnancy and childbirth on their bodies, particularly in locations with minimal healthcare facilities. Lower-income women and women who are marginalized (for a variety of factors, such as race, ethnicity, class, or political or religious affiliations) are at greater risk of pregnancy complications, including death. Other maternal health concerns include women's risk for anemia, preeclampsia, eclampsia, malaria and infections. Changing climates and climatic variations have also affected maternal health; for example, malaria and anemia were found to be linked to seasonal variations corresponding with rainfall in sub-Saharan Africa but not in South or Central Asia, while eclampsia was more often linked to colder temperatures (Hlimi 2015). Geographic differences, particularly between urban centers, can be another reason for disparities in health outcomes, particularly when healthcare services remain minimal in rural areas with limited midwifery programs (Hotchkiss 2001).

Much of the research on maternal health further identifies the role of gender inequality and improving women's decision-making ability within societies as a health issue. For example, when women have greater decision-making capabilities, they are more likely to receive prenatal and **antenatal care** by birth attendants, midwives, or obstetrics and gynecology doctors (Adjiwanou and LeGrand 2014). Gender inequality is a health issue, particularly when women lack the ability to make decisions about their health and healthcare needs. In some places, women may experience sexism within the healthcare system. For example, in Uganda, while women are the primary users of healthcare services, they have experienced exclusions within healthcare structures based on sexist barriers. Discrimination against women within the health sector included sexist or inappropriate remarks, which caused women to avoid healthcare centers for themselves and their children (MacKian 2008, 110).

Maternal mortality has become an issue of concern in the United States, which has much higher maternal mortality rates compared to other high-income countries (Kassebaum and Maternal Mortality Collaborators 2016). Maternal mortality rates in the United States more than doubled between 1990 and 2013. An estimated 1,200 women suffer from childbirth or pregnancy difficulties in the United States. Chronic conditions such as hypertension, obesity, and diabetes contribute to pregnancy-related complications. Racial and economic inequalities are also major factors; for example, pregnant women without health insurance are three to four times more likely to die than those with insurance, and African American women are three to four times more likely to die of pregnancy-related complications than white women (Main and Menard 2013, CDC 2017).

Intersectional inequality remains a key factor in health determinants for women and men, which is further evidenced by the identification and treatment of sexually transmitted diseases such as HIV/AIDS. When women have limited ability to control the conditions of their sexual relationships along with unequal access to healthcare provisions, their risk of contracting sexually transmitted diseases increases (Aveling 2012). Understanding gender roles and relations within the sociocultural and spatial contexts within which they exist is vital for effectively treating individuals and preventing the spread of disease. In many places throughout the globe, HIV/AIDS affects men at a much higher rate than women; however, in sub-Saharan Africa women are more likely than men to contract HIV/AIDS (Wyrod 2016). Research on gender and HIV/AIDS in Africa identifies the need to examine masculinity in order to understand the gendered characteristic of AIDS, "especially how men and women navigate varying ideals of male sexuality" (Wyrod 2016, 17). Thus, these studies show how social science research reveals social identities and marginalization based on gender, race, sexuality, class and dis/ability. Social science research provides a better understanding of why certain health disparities exist and can help in shaping how to mitigate both health (and social) inequalities.

While a significant amount of attention and funding has been allocated to treat and prevent the spread of HIV/AIDS globally, several issues related to gender roles/relations and disparate conceptualizations of acceptable sexual behavior must also be taken into consideration. For example, in places where homosexuality is forbidden, taboo or legally restricted, men who have sex with men may be less likely to seek treatment for sexually transmitted infections or admit to having homosexual sex because they fear social or political censure, condemnation, arrest or abuse. In other contexts, local health practitioners adopt gendered healthcare frameworks that will appeal to donors' interests rather than the recommendations of local healthcare providers or the gendered needs of communities (Mannell 2014). Thus, gender, development and health scholars identify the importance of addressing fundamental gender inequalities in order to improve women's health. Political influences on women's reproductive health must also be considered.

Political interest in public health includes addressing infectious and contagious diseases along with issues of population growth. Population growth can be encouraged to ensure younger generations produce enough children to eventually replace the aging population. Population control measures are conversely used to mitigate population increase. Some countries have identified population control efforts as necessary for improving public health, decreasing maternal and infant mortality, and deceasing poverty. However, closer examination of these measures demonstrates that many population control programs are uneven, i.e., only decreasing the reproduction of specific groups of people based on racist and classist assumptions about them. Gender, race, sexuality and economic inequality remain major contributors to health disparities such as the ability (or lack thereof) of individuals to access nutritious and affordable food, clean water, healthy and safe living and working environments, and quality healthcare.

Governments and supranational organizations that attempt to control populations seek to prevent women and men from choosing their own reproductive paths. Development organizations have provided women (and men) with birth control in an effort to decrease the number of births in certain locations. However, several of the family planning projects sponsored by the IMF and World Bank included abuse, classism, gender-biased and racist approaches to curbing population growth (Wangari 2002). While many women have benefited from the ability to access birth control, others – particularly poor women – have experienced negative effects. Therefore, while the availability of birth control can be positive when women are able to choose if, when and how many children they have, these choices are often mitigated by the socioeconomic and cultural context within which they live, as individual choice is not always possible or desirable for some women.

Population growth and management

Many governments, for a variety of reasons, are invested in managing the population within their borders. This section examines some basic ideas of demography and how they relate to gender and population management. Birth and death rates are used to identify or understand population growth. **Birth rates** are calculated by the number of live births per 1,000 persons, and **death rates** are calculated by the number of deaths per 1,000 persons. The **rate of natural increase** (RNI) for a population is calculated by subtracting the death rate from the birth rate. Rates of **infant** and **maternal mortality** (how many infants die before one year of age, and how many women die in childbirth) as well as **total fertility rates** (how many children a woman has in her lifetime, regardless of infant mortality) are also strong indicators of reproductive health and population change. **Life expectancy** indicates the average age of death for persons within a particular country. This measure identifies global disparities across countries, while social and economic inequalities reveal sharp differences in the length of one's life within a country. Macro-scale quantitative data provides useful information to gauge differences in life expectancy (and other measures across countries), while social science research and qualitative analyses provide detailed and nuanced information that uncover the reasons why these disparities persist.

Population pyramids are helpful for illustrating population growth or decline within a specific country based on age and sex. A population pyramid provides an overall graphic illustration of age distributions separated by gender for a given time in the population's history. These pyramids help researchers to visualize a population's growth or decline. For example, Japan's population pyramid reveals an aging population, declining birth rates and relatively high life expectancy (see Figure 7.2), while Sierra Leone's population pyramid indicates a much younger population, higher birth rates and lower life expectancy (see Figure 7.3). These types of graphs are useful for describing different population patterns, while social science research methods are helpful for explaining *why* these patterns exist.

Several countries have linked population management with public healthcare. For example, some governments use a portion of collected taxes to provide citizens with free access to healthcare. Government-generated healthcare programs often focus on public health issues, mitigating the spread of disease, and regulating populations. Governmental, nongovernmental and supranational organizations, such as the World Health Organization (WHO), use various measures to classify the health of a population. While these measures remain prominent for quantifying the general health of a population, the calculations do not include a more qualitative understanding of a population's health and healthcare needs. Examples of qualitative measures include: understanding social networks, family assistance, social capital, friendship networks and assistance, contentment, happiness and caregiving. Attempts to consider mental health as part of an overall health profile have begun to address these less noticeable influences on one's life, while quantifiable measures predominate in attempting to calculate the health of a population, especially at a macro scale. Global development and health calculations further

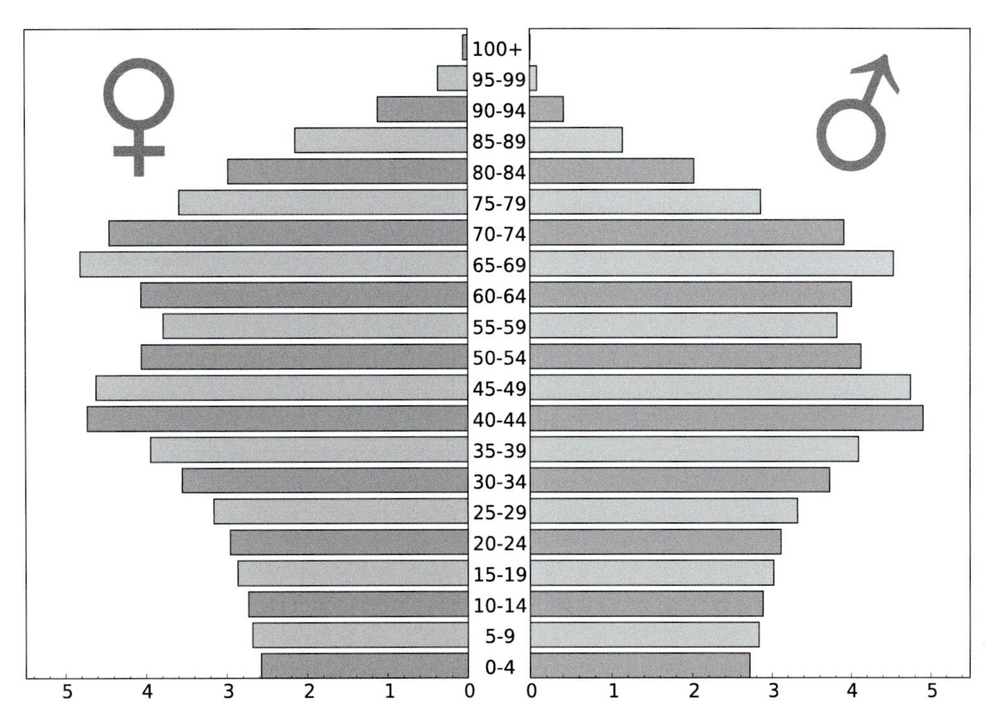

Figure 7.2 Japan's population pyramid

Source: by MagHoxpox [CC BY-SA 4.0 (https://creativecommons.org/licenses/by-sa/4.0)], from Wikimedia Commons

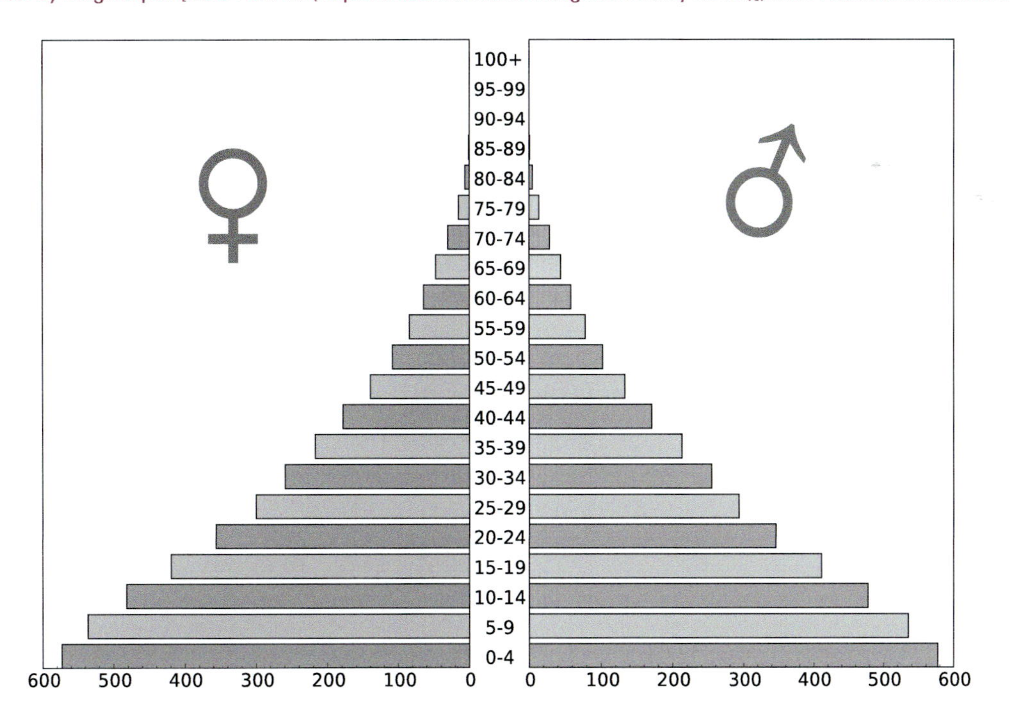

Figure 7.3 Sierra Leone's population pyramid

Source: by MagHoxpox [CC BY-SA 4.0 (https://creativecommons.org/licenses/by-sa/4.0)], from Wikimedia Commons

attend to population growth and control, which are based on divergent ideas about the health of people throughout the world.

Gender has become an essential organizational category of analysis for improving health outcomes. In 2007 the WHO adopted a strategy for "integrating gender analysis and actions" into its work, which remains part of its overall mission. The WHO, along with other health-focused development organizations, recognizes the importance of integrating a clear understanding of how certain health risks are gendered and require different forms of treatment. Gender-inclusive and sensitive healthcare provisions try to attend to men's and women's specific health needs, such as sexual and reproductive health. Women remain the primary group targeted for programs that seek to address the health needs of children, because women, in general, remain the primary caregivers of newborns and young children. Therefore, ensuring that primary caregivers (the majority of whom are women) are healthy and have access to healthcare has a positive effect on the health of the children and families they care for. The availability of healthcare for women and men is often contingent on economic resources, both in terms of seeking and being able to access care. The cost of healthcare remains a central challenge for individuals across the globe.

Population concerns

While population control measures continue to be implemented by states and supranational development organizations, most geographers and related social science scholars identify resource allocation, consumption patterns and growing inequalities between rich and poor as far greater concerns than population numbers and growth, particularly with respect to resource allocation and environmental degradation (Sheppard et al. 2009). Many contemporary population control measures are based on flawed neo-Malthusian logic. Malthus was an 18th-century demographer who believed that population growth would eventually exceed the ability of agricultural production to supply enough food and resources for the global population. Subsequently, Malthusian logic promoted population control as a method for ensuring enough resources. Neo-Malthusians advocate for the use of contraception and sterilization in an effort to curb population growth. Malthus's logic was flawed because he did not address massive changes in wage labor practices, which expected rural peasants to migrate to urban spaces and led to the increased alienation of these individuals from subsistence food production. Once people are required to work for food and denied access to subsistence agriculture, hunger increases dramatically (Neeson 1996).

In the 20th century, Danish economist Ester Boserup (1965, 1970) challenged Malthus's logic by showing that population growth guided the amount of agricultural production needed to meet the food demands of a given population and therefore human innovation attends to changing resource needs. Boserup's research identified the ways in which populations found new ways and methods for increasing food production and resources as populations grow, but did not address the fundamental problem of political and economic control over natural resources. Critics of Boserup identify her lack of political analysis, such as not examining the state's role in economic and food systems, market relations and the misuse of surplus resources (Sheppard et al. 2009). For example, several forms of farming innovation, such as the use of pesticides, have increased crop yield, while introducing carcinogens into the environment, and endocrine-disrupting chemicals, which have been linked to obesity (Guthman 2011). Additionally, the use agricultural "innovation" within a capitalist framework has overwhelmingly benefited corporations and their shareholders.

Both Malthus's and Boserup's interventions included strategic flaws. Malthus was flawed because population growth in and of itself does not lead to lack of resources. Boserup's ideas that privileged human ingenuity were also shortsighted because they did not incorporate politics and the pitfalls of human innovation, particularly when it privileges profits over people. Paid water services, pesticide use (the Green Revolution) and GM crops exemplify the manipulation of human innovation and technology that increasingly places environmental resources under the control of the wealthy and powerful. Therefore, it is imperative to examine the opportunities, pitfalls and challenges of human innovation and technology (see Chapter 8).

Geographers have also shown how food disparities, such as famine, have been created from the political economy of competing countries and empires rather than on agricultural limitations or need for innovation (Davis 2002). Therefore, famines and food scarcity have been used as political weapons. Today population growth measures continue to influence state policies and international development programs, while the growing divisions between rich and poor become increasingly part of mainstream discourse. Some population control measures have been implemented under the banner of ensuring public health and access to resources, while in practice they often perpetuate racist, classist and xenophobic ideologies, which are implemented through intersectional forms of gender discrimination. It is important to note that there are significant differences between birth control and population management. Birth control is a method of family planning that allows individuals, couples or families to decide if, when and how many children to have. Conversely, population control measures use reproductive technologies to regulate the number of children born in a country or among a particular group of individuals.

Population control

Population growth remains a development issue; however, the global population growth rate is approximately 1.14 percent per year, the lowest growth rate since early in the 20th century, and global fertility is "currently 2.3, down from 4.95 in 1950" (Robbins and Smith 2017, 200). Robbins and Smith (2017) identify women's increased participation in education, household decision making, employment and access to birth control as major reasons for the rapid decline in global birth rates. This research identifies some of the successes of gender-based development programs, while also highlighting how powerful entities inflate the birth rates of certain groups in an effort to limit their reproduction and further marginalize them socially and politically. In spite of the effectiveness of contraceptive technologies in reducing unwanted pregnancies, abortion rates and population growth, many people's contraception needs remain unmet. Thus, critically examining reproductive health measures illustrates how they have been used as tools for allowing and encouraging the reproduction of some groups, while limiting the birth of others.

Some governments politically manipulate reproductive technologies in order to control the number of births for certain women (based on their race, class, mental and physical capabilities, or sexuality) and to decide which individuals/groups can and cannot reproduce. Historically, women's individual choices to use contraception or terminate a pregnancy were criminalized in various countries, including the United States. For example, Margaret Sanger, a birth control advocate in the early 20th century, fled the United States to avoid criminal prosecution. In an effort to legitimize the development, improvement and use of contraception, she joined the eugenics movement. Eugenics is the practice of using birth control technologies to genetically alter a society. Eugenic practices use reproductive technologies to increase sexual reproduction among people with "desirable" traits and reduce or stop the rates of reproduction by way of contraception and sterilization of individuals identified as having "undesirable" traits. Eugenics is inherently a racist and malicious project as it favors one group over another and those in positions of power determine the reproductive practices for certain groups within a population without their input or permission.

Who gets to decide which humans reproduce and which do not? Controlling reproduction further illuminates various forms of uneven power and structural violence associated with various population control measures and methods. Eugenic uses of reproductive technologies are mainly for the purpose of selecting who those in power identify as "the best humans." The racist belief in "better humans" reinforces existing social hierarchies and divisions between groups based on the assumption that people can be biologically engineered to meet the ideals defined by the powerful who subscribe to their own sense of self-importance and superiority.

Most individuals involved in the global eugenics movement sought to limit or prevent the reproduction

FOCUS: GENDER, RACE AND POPULATION CONTROL

China's "one-child policy" exemplifies one of the largest efforts by a country to curb population growth. This policy was initiated in 1979 and China's State Family Planning Bureau set the benchmarks and implementation plans for the program. Parents who exceeded the limit of children were subject to high fines and their children were excluded from free public education. However, the one-child rule was not applied to the population uniformly. For example, in rural areas a second child was allowed after five years if the first child born was a girl. Son preference has been pervasive in rural China, because sons are expected to stay with their birth family and assist with agrarian production and aging parents, while daughters are expected to marry into another family. Therefore, having a daughter was less desirable for some parents. Son preference led to sex-selective abortions of female fetuses, which created sex-ratio imbalances at birth; for example, in 2005, 118 boys were born for every 100 girls (Ebenstein 2010). The one-child policy in China has become less restrictive in recent years. In 2015, the Chinese government announced that couples would be allowed to have two children.

of people considered less desirable based on racist and ablest assumptions about human progress. Eugenics became an integral part of Nazi Germany, while similar programs were occurring in the United States at the same time. Power is expressed through the ways in which eugenics (sometimes renamed as **social biology**) has been and continues to be used. The actions of individuals in positions of power, authority or self-identified superiority must be questioned and resisted, particularly when they are based on an arbitrary set of standards for categorizing individuals through a perverse hierarchical ordering of corporeal, economic or social traits. Racism is built upon these divisions and categorizations, which increases intersectional inequalities. Population control measures continue to this day through various development projects and the spread of reproductive technologies. For example, China's "one-child policy" limited the number of children that could be born per family (see the focus section, above).

The Israel–Palestine conflict has included battles over reproduction. Mayer (2012) uses the term "womb wars" to identify efforts by the Israeli government to encourage Jewish women to have many children, while attempting to limit births among Palestinian women. In another case, the rise of Hindu nationalism in India manifested in purposeful changes in marriage and reproductive practices (Smith 2012). Political discourse or policies that shape marriage or reproductive practices among individuals or groups are referred to as **intimate geopolitics** (Smith 2012). For example, within the Leh, Ladakh, community of northern India, Muslims and Buddhists had openly intermarried without incident until relatively recently. Since the rise of Hindu nationalism and the marginalization of Muslims in various areas within India, intermarriage has become increasingly taboo between these groups (Smith 2012). These concerns have included popular misconceptions about the growth of the Muslim population based on incorrect assumptions about the size and birth rates of Muslim families. The following two case studies further examine the politics of gender, health, population decline and growth over time in the Czech Republic and India.

Case studies: Czech Republic and India

The Czech Republic's changing gender roles from socialism to capitalism are examined in relation to efforts to incorporate women into the labor force while ensuring population growth. Conversely, concerns about the size of India's population were historically viewed as a public health issue, leading to eugenic-inspired population control measures. These efforts did little to effectively mitigate the effects of population growth or address public health concerns, which were promoted as the basis for these extreme measures.

Czech Republic

The Czech Republic's population is estimated at over ten million. Population growth was slow in the mid-20th century due to the loss of population during World Wars I and II and later in the 21st century due to low fertility rates, which were caused by women choosing to have fewer children and delaying pregnancy. Like many European countries, the Czech Republic is faced with the demographic problem of a declining population. The Czech population was at its height of just over 11 million at the beginning of World War II. However, loss of life and the expulsion of the German population from the country after World War II (in response to Nazi occupation during the war) contributed to the decline of the population to 8.8 million. Population growth resumed and by 1994 the population had increased to 10.33 million. Natural growth was negative between 1994 and 2003 (–0.15 percent per year). Since 2005, growth has been more positive, with the most recent increases to population coming from an influx of immigrants. By focusing on changing gender roles in the former Czechoslovakia under socialist regimes, the following provides a gendered overview of the economics and politics of population growth and decline from the dissolution of Czechoslovakia into the Czech and Slovak republics, and the transition from socialism to capitalist development.

Soon after World War II, Czechoslovakia was ruled by the communist party, and state socialism was shaped by gender ideologies that often conflicted with official doctrines of communism and central planning, which viewed women as integral to productive state labor (Wagnerová 2016). The socialist regime did not effectively address the gendered methods of social reproduction (i.e., women as homemakers and caregivers) because it privileged the increased development of industrial production over everyday human consumption needs. Therefore, the Czechoslovakian government did not recognize or address existing gender relations and gender divisions of labor that associated domestic reproductive labor (inside the home) with women, and productive labor (outside the home) with men.

Women experienced double time-to-labor burdens under the socialist regime, which encouraged them to work outside the home but did not encourage men to assist with women's existing domestic labor practices. This resulted in much greater gender inequalities, particularly in work-time compensation. "Ironically, the policies of the socialist state reinforced the gendered distinction between public and private, and the liberal valorization of the individual family over the collective . . . while officially denouncing them" (True 2012, 39). Interestingly, men's economic power was historically tied to patriarchal domination; therefore, the nationalization of wealth under socialism was an economic loss for men, similar to the emergence of private property being an economic loss for women because of their marginalization from productive labor in many sectors (Jusová and Šiklová 2016). Both men and women experienced labor exploitation under socialism with significant negative impacts on their health, particularly upper respiratory illnesses due to poor ventilation in government factories. Healthcare was provided by the state but accessing care in a timely manner remained difficult for many. Additionally, public health initiatives associated with population growth and birth control focused on issues of abortion and contraception.

Prior to the onset of socialism, Czechoslovakia was a Catholic country and staunchly opposed to both contraception and abortion. Under socialism the church was politically marginalized, paving the way for leftist women's groups to advocate for abortion rights. Reproductive rights for women under socialism were heavily debated, especially between 1955 and 1956, with primary arguments focusing on the dignity of socialist women not having to resort to illegal "back alley" abortion methods. Partly due to the lack of anti-abortion protestors, abortion was legalized in 1958; however, women were required to receive approval from a local abortion committee. Despite worries about these committees blocking abortions, they approved 90 percent of requests. The local abortion approval committees were abolished in 1986. The highest number of abortions within the country occurred 1968–69, and 1988–89, which marked two eras of significant political upheaval (the Soviet Occupation and the Velvet Revolution, respectively), suggesting that increased abortions were in response to political uncertainty and insecurity.

In the 1980s the birth rate in Czechoslovakia began declining, leading to national concerns about sustaining and replacing the population. In the 1990s both the birth rate and abortion rate declined. This decade was also marked by a major economic transition from socialism to capitalism and the dissolution of Czechoslovakia into two separate, independent states: the Czech and Slovak republics. During this transitional phase, maternal health and childbirth transitioned from the use of midwives and home births to the expectation that women would give birth in hospitals. Public debates situated midwives as risk-taking independents, and mothers who used midwives as irresponsible. Conversely, maternity ward doctors were identified as "the rational, reasoned, benevolent 'masculine' experts who promised maximum safety for child delivery in fully equipped maternity wards" (Šmídová 2016). This transition illustrates the ways in which the medicalization of women's health in capitalist economies mirrors women's lack of access to decision making in other sociopolitical contexts targeted for economic development.

The transition from socialism to capitalism proved difficult for men in many former communist states. Health concerns associated with capitalism include the use and abuse of alcohol, tobacco and marijuana throughout the Czech Republic, particularly among young people (Spilkova, Dzúrova and Pitonak 2014). The consumption of alcohol and tobacco in the Czech Republic is both ubiquitous and accepted; therefore, the government has attempted to curb these behaviors in public spaces. Alcohol, tobacco and drug consumption have therefore become a public health issue with programs designed to curb individual behaviors for the public good. For example, in May 2017, smoking in public places (restaurants, bars, cinemas and sporting events) was banned throughout the country.

Czech men's and women's labor was exploited under communism; however, men did not (and were not expected to) participate in unpaid household labor. Men's experiences of household work, being laid off or demoted occurred as part of the transition from communism to capitalism. Women's participation in the new capitalist labor market is high, and women are choosing to delay when they have children and not having as many children as the generation before them. Birth rates continue to decline as many contemporary women push back against the double burden of working in and outside the home (True 2012).

While communist states such as Czechoslovakia may have discursively supported women's emancipation from patriarchy, the actual practice of women's empowerment through participation in the labor force had the added effect of increasing women's time-to-labor burdens. Currently there are concerted efforts by the Czech government to encourage women to have children by providing state-sponsored maternity leave policies. For example, the government pays up to 28 weeks of maternity leave at 70 percent salary for women, and up to 156 weeks of paid parental leave (for either parent) at a flat rate. Specific paternity leave was not provided until recently. It now provides up to seven days' paid leave at 70 percent salary to fathers within the first six weeks of their child's birth. Therefore, in an effort to increase birth rates among Czech women, the government provides paid maternity leave in order to encourage women to have children without severely sacrificing their participation in the paid labor force.

India

In contrast to the Czech Republic's declining population, India has experienced extensive population growth throughout the 20th century and into the 21st century. Historically, the All India Women's Conference (AIWC) advocated for female access to birth control beginning in 1932 in an effort to improve women's health. The AIWC supported birth control for hygienic, economic and eugenic reasons.

Margaret Sanger, a US-based birth control advocate and eugenicist, assisted with birth control advocacy in India, focusing mainly on improving maternal health and decreasing maternal and infant mortality rates. During her tour of India, Sanger sought to demonstrate a direct link between birth rates and infant and maternal mortality (Prasad 2007). Controlling the population was further linked to deceasing poverty and improving prosperity for Indian citizens. Birth control methods continued but remained controversial in India, until the widespread government effort to curb population growth in the 1970s. From 1971 to 1973 vasectomy camps were set up first in Kerala and then spread throughout the country.

Male sterilization numbered 1.3 million in 1970–71 and by 1973 had grown to 3.1 million. The Minister of State for Family Planning in India suggested compulsory population control measures through male sterilization programs, but this policy was highly controversial and not implemented until 1975. Prime Minister Indira Gandhi declared a state of emergency in India in 1975, which suspended the constitution as well as the civil rights and liberties of Indian citizens. Gandhi used this time to suppress political uprisings and jail her opponents, and massive population control measures were initiated and orchestrated with the assistance of her son, Sanjay Gandhi. The declaration of the state emergency accelerated existing trends toward population control in India. The World Bank and IMF used this opportunity to influence Indian politics by supporting and pushing for radical population control measures to alleviate poverty in India. Sanjay Gandhi spearheaded population control measures – including forced and coercive sterilization programs – in an effort to curb population growth and appease Western development donors.

Throughout India both positive and negative efforts were initiated to curb population growth. Positive incentives included pay rises for government workers who underwent sterilization. On the negative side, some states withdrew maternity leave benefits for female employees after two births. Forced or coerced sterilization efforts (such as vasectomy and tubal ligations) continued during this period. Approximately 8.25 million people were sterilized between 1976 and 1977 – 6.2 million were men, and more than 2,000 died from complications or botched operations. While these programs were widespread they did little to alleviate poverty in India.

The national sterilization programs in India came under severe criticism and scorn both within the country and internationally.

> In the heat of pursuing the intermediate objective of improved family planning performance, the pursuers lost sight of the difference between family-planning performance as a goal and the larger end of human well-being to which fertility reduction had originally been meant to contribute.
>
> (Gwatkin 1979, 51)

Additionally, using population control to decrease poverty and improve public health was never achieved through this program. India's population growth continues to be a topic of debate today. However, recent government efforts to curb population growth have taken a softer approach through public service announcements, billboards and advertisements that suggest two children as the optimal number for any family (Prasad 2007).

The common phrase "A small family is a happy family" has become ubiquitous throughout India, including being printed on a postage stamp with a picture of two parents and two children (Prasad 2007). The small family/happy family campaign identifies the "ideal" Indian family as nuclear – two heterosexual parents and two children (usually pictured as one girl and one boy). Despite this use of soft power through public service announcements and campaigns, sterilization camps (mostly for women) continue to operate in some locations within India. These camps do not provide adequate care for women and recent investigative journalist reports have uncovered serious injuries and deaths among the women being sterilized. Boy preference among some patriarchally structured families in India transformed into negative birth control measures such as sex-selective abortions. The extensive increase in sex-selective abortions, particularly in northern Indian states, has led to

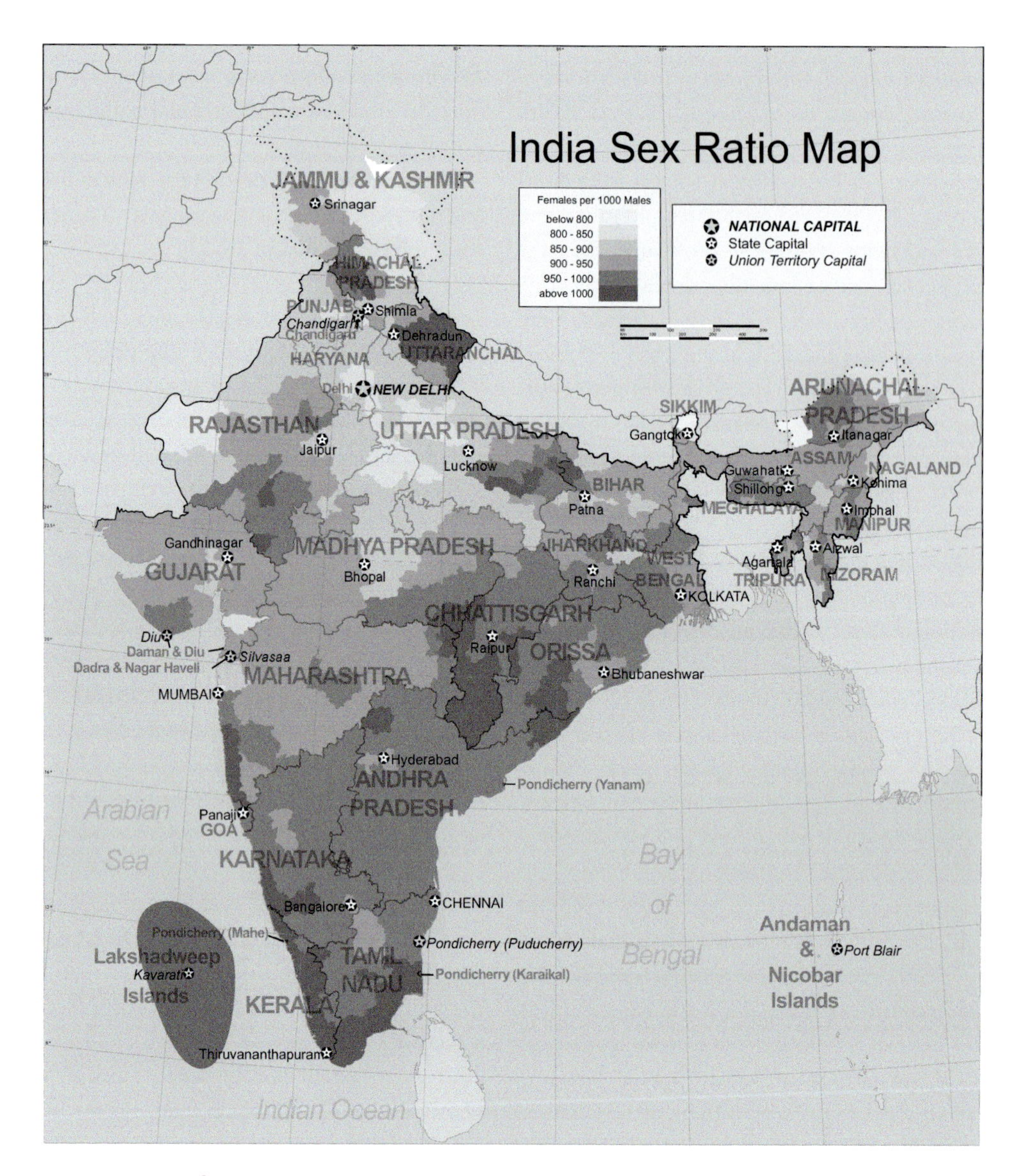

Figure 7.4 2011 map of Indian states sex ratio

Source: by Planemad [CC BY-SA 3.0 (https://creativecommons.org/licenses/by-sa/3.0)], from Wikimedia Commons

a significant imbalance in male to female birth rates (see Figure 7.4). In response to this extensive skewing of gendered birth rates in northern India, the government made it illegal for clinics or hospitals to reveal the sex of a fetus prior to the birth of a child.

Extensive divisions between the extremely wealthy and poor have continued to grow since the 1990s when India began to open its economy to neoliberal global capitalism. While the Indian government continues to provide welfare and social services to the poor, the provision of these services is often slow and stymied by bureaucracy and corruption (Gupta 2012). Transnational surrogacy is one of the ways in which global capitalist-based corporations have integrated health tourism with biological reproduction. International surrogacy companies have seized upon

Figure 7.5 A small family is a happy family – postage stamp

Source: neftali/Shutterstock.com

the economic needs of poor women in India (and eight other countries) to allow couples to implant an in-vitro-fertilized egg into the womb of an Indian surrogate. Wealthier women and couples who cannot have children seek surrogacy in order to artificially inseminate their fertilized egg into the body of a woman who is paid to carry the fetus and give birth to the child (see Chapter 8). Therefore, reproductive technologies often highlight national and global economic divisions and the use and abuse of power by governments.

Future challenges and opportunities

Individual and public health continues to be integral to many development programs globally. As discussed in several studies, addressing gender inequalities and increasing women's empowerment improves women's access to healthcare. When women and men have the ability to make their own decisions about their healthcare, they are better able to ensure the best care for themselves and their families. Living in a healthy environment and having quality healthcare remain limited for many individuals, particularly when one's ability to access healthy food, water, air or care is contingent upon one's social, racial, economic or political class/status.

Population control measures, as discussed in this chapter, are often related to improving public health and access to resources. However, the implementation of these programs in many countries has been racist, sexist/heterosexist, classist and xenophobic. Additionally, economic development projects that seek to curb populations divert attention away from areas of low population growth but high mass consumption. Thus, curbing consumption has a much greater potential to improve the health of environments globally (and mitigating climate change) than controlling population growth among the world's poor.

Universal healthcare is one method for addressing economic inequalities in healthcare access. Governments that provide basic healthcare services for all residents can improve individual access to doctors and hospitals along with preventive care. For example, when governments are responsible for the health and healthcare of their citizens, they have an economic, social and political imperative to ensure healthy environments. This leads to improved government regulations on industries that pollute environments and compromise the health of many people. However, ineffective bureaucracies are similarly problematic, as people can quite literally die while waiting to receive healthcare services (Gupta 2012). Therefore, improving access to healthy foods, environments and workspaces must occur in tandem with providing affordable and quality healthcare.

Recommended reading

The handmaid's tale, Margaret Atwood; *Living downstream*, Sandra Steingraber; *Having faith*, Sandra Steingraber; *Baby markets: money and the new politics of creating families*, Michele Bratcher Goodwin (Ed.)

Recommended viewing

Something like war; *Clothes to die for*

Questions for discussion

What forms of population control continue to exist today? What are some of the most pressing public health issues in your community and around the globe? What are the positives and negatives associated with government-based universal healthcare programs? What are the positives and negatives associated with capitalist-based healthcare programs, such as for-profit health insurance providers?

References

Adjiwanou, V., & LeGrand, T. (2014). Gender inequality and the use of maternal healthcare services in rural sub-Saharan Africa. *Health & Place*, 29, 67–78.

Aveling, E.-L. (2012). Making sense of "gender": from global HIV/AIDS strategy to the local Cambodian ground. *Health & Place*, 18(3), 461–67.

Boserup, E. (1965). *The conditions of agricultural growth*. Chicago, IL: Aldine.

Boserup, E. (1970). *Women's roles in economic development*. New York, NY: St. Martin's Press.

Buor, D. (2004). Water needs and women's health in the Kumasi metropolitan area, Ghana. *Health & Place*, 10(1), 85–103.

CDC (2017). Meeting the challenges of measuring and preventing maternal mortality in the United States. Centers for Disease Control. www.cdc.gov/grand-rounds/pp/2017/20171114-maternal-mortality.html. Accessed 2/13/2019.

Cleaver, F. (Ed.) (2002). *Masculinities matter! Men, gender and development*. London and New York: Zed Books.

Davidson, J., & Stratford, E. (2007). En(gender)ing the debate about water's management and care – views from the Antipodes. *Geoforum*, 38, 815–27.

Davis, M. (2002). *Late Victorian Holocausts: El Nino famines and the making of the Third World*. London: Verso.

Ebenstein, A. (2010). The "missing girls" of China and the unintended consequences of the one child policy. *Journal of Human Resources*, 45(1), 87–115.

Gupta, A. (2012). *Red tape: bureaucracy, structural violence, and poverty in India*. Durham, NC: Duke University Press.

Guthman, J. (2011). *Weighing in: obesity, food justice, and the limits of capitalism*. Berkeley, CA: University of California Press.

Gwatkin, D.R. (1979). Political will and family planning: the implications of India's Emergency Experience. *Population and Development Review*, 5(1), 29–59.

Halvorson, S.J. (2004). Women's management of the household health environment: responding to childhood diarrhea in the northern areas, Pakistan. *Health & Place*, 10(1), 43–58.

Hlimi, T. (2015). Association of anemia, pre-eclampsia and eclampsia with seasonality: a realist systematic review. *Health & Place*, 31, 180–92. doi:10.1016/j.healthplace.2014.12.003.

Hotchkiss, D.R. (2001). Expansion of rural health care and the use of maternal services in Nepal. *Health & Place*, 7(1), 39–45.

Jusová, I., & Šiklová, J. (Eds) (2016). *Czech feminisms: perspectives on gender in Eastern and Central Europe*. Bloomington, IN: Indiana University Press.

Kassebaum, N.J., & Maternal Mortality Collaborators (2016). Global, regional, and national levels of maternal mortality, 1990–2015: a systematic analysis for the Global Burden of Disease Study 2015. *The Lancet*, 388, 1775–1812.

MacKian, S.C. (2008). What the papers say: reading therapeutic landscapes of women's health and empowerment in Uganda. *Health & Place*, 14(1), 106–15.

Main, E.K., & Menard, M.K. (2013). Maternal mortality: time for national action. *Obstetrics & Gynecology*, 122(4), 735–36.

Mannell, J. (2014). Adopting, manipulating, transforming: tactics used by gender practitioners in South African NGOs to translate international gender policies into local practice. *Health & Place*, 30, 4–12.

Mayer, T. (2012). The struggle over boundary and memory: nation, borders, and gender in Jewish Israel. *Journal of International Women's Studies*, 13(4), 29–50.

Neeson, J.M. (1996). *Commoners: common right, enclosure and social change in England, 1700–1820*. Cambridge: Cambridge University Press.

O'Reilly, K., Halvorson, S., Sultana, F., & Laurie, N. (2009). Introduction: global perspectives on gender – water geographies. *Gender, Place and Culture*, 16(4), 381–85.

O'Reilly, K., & Louis, E. (2014). The toilet tripod: understanding successful sanitation in rural India. *Health & Place*, 29, 43–51.

Prasad, S. (2007). "A small family is a happy family: the politics of population control in India." In K.K. Misra, & J.H. Lowry (Eds) *Recent Studies on Indian Women*, Jaipur: Rawat (pp. 253–80).

Pratt, G. (2012). *Families apart: migrant mothers and the conflicts of labor and love*. Minneapolis, MN: University of Minnesota Press.

Reed, M.G., & Christie, S. (2009). Environment geography: we're not quite home – reviewing the gender gap. *Progress in Human Geography*, 33(2), 246–55.

Robbins, P., & Smith, S.H. (2017). Baby bust: towards political demography. *Progress in Human Geography*, 41(2), 199–219.

Sabhlok, A., Cheung, H., & Mishra, Y. (2015). Narratives of health and well-being: migrant road workers in the upper Himalayas. *Economic & Political Weekly*, 50(51), 71–78.

Schenker, M.B. (2010). A global perspective of migration and occupational health. *American Journal of Industrial Medicine*, 53(4), 329–37.

Sheppard, E., Porter, P.W., Faust, D.R., & Nagar, R. (2009). *A world of difference: encountering and contesting development*. New York, NY: Guilford Press.

Silvey, R. (2004). Transnational domestication: state power and Indonesian migrant women in Saudi Arabia. *Political Geography*, 23(3), 245–64.

Šmídová, I. (2016). "Condemned to rule: masculine domination and hegemonic mascuilinities of doctors in Czech maternity wards." In I. Jusová, & J. Šiklová (Eds) *Czech feminisms: perspectives on gender in Eastern and Central Europe*, Bloomington, IN: Indiana University Press (pp. 222–36).

Smith, S. (2012). Intimate geopolitics: religion, marriage, and reproductive bodies in Leh, Ladakh. *Annals of the Association of American Geographers*, 102(6), 1511–28.

Sparke, M. (2013). *Introducing globalization: ties, tensions, and uneven integration*. Sussex: Wiley-Blackwell.

Spilkova, J., Dzúrova, D., & Pitonak, M. (2014). Perception of neighborhood environment and health risk behaviors in Prague's teenagers: a pilot study in a post-communist city. *International Journal of Health Geographics*, 13(1), 41.

Stoler, J., Fink, G., Weeks, J.R., Otoo, R.A., Ampofo, J.A., & Hill, A.G. (2012). When urban taps run dry: sachet water consumption and health effects in low income neighborhoods of Accra, Ghana. *Health & Place*, 18(2), 250–62.

Sultana, F. (2011). Suffering for water, suffering from water: emotional geographies of resource access, control and conflict. *Geoforum*, 42(2), 163–72.

True, J. (2012). *Gender, globalization, and postsocialism: the Czech Republic after communism*. New York, NY: Columbia University Press.

Wagnerová, A. (2016). "Women as the object and subject of the socialist form of women's empancipation." In I. Jusová, & J. Šiklová (Eds) *Czech feminisms: perspectives on gender in Eastern and Central Europe*, Bloomington, IN: Indiana University Press (pp. 77–94).

Wangari, E. (2002). "Reproductive technologies: a Third World feminist perspective." In K. Saunders (Ed.) *Feminist post-development thought: rethinking modernity, post-colonialism, and representation*, London and New York: Zed Books (pp. 298–312).

WHO (2017). Human rights and health. www.who.int/en/news-room/fact-sheets/detail/human-rights-and-health. Accessed 2/17/2019.

Williams, G., & Mawdsley, E. (2006). Postcolonial environmental justice: government and governance in India. *Geoforum*, 37(5), 660–70.

Wyrod, R. (2016). *Aids and masculinity in the African city*. Berkeley, CA: University of California Press.

8 Gender and development technologies

Introduction

Technology transfer has been a key factor in economic development globally. The idea of transferring technological skills from countries with extensive technological proficiency and production to countries without or with less technological advantage continues to this day. The history of technology transfer is extensive, dating to ancient trade routes and later being an integral part of Western European colonization. Colonial powers viewed technology as an advantage, which was used to extract resources, wealth and work from colonized places and people. Postcolonial development after World War II continued to use technology as a marker of progress. Technological prowess became integral to development efforts led by the United States and Soviet Union after World War II. These projects took form as large-scale, centralized infrastructure projects, such as dam construction, road building and agricultural mechanization, including the manufacture and use of pesticides, urban infrastructure such as electricity, along with water and sanitation improvements. Factory construction and automation in several industries followed these trends. More recently, technology has included a massive increase in robotic automation and communication technologies, from cell phones to social media. Technology has also been viewed as an overarching "fix" or method of improving societies, particularly in countries considered un- or under-developed. In this chapter, we examine the concept of technological transfer, economic development and intersectional gender experiences through three economic sectors: 1) labor and manufacturing; 2) information technologies; and 3) biotechnologies.

We begin with a general overview of the relationship between gender and technology, highlighting male dominance in the creation of technologies, particularly to improve infrastructure and ensure the flow of capitalist production and consumption. This is followed by a discussion of how capitalist development and globalization have influenced these three sectors of gendered work. We conclude with two case studies, one that highlights the gendered use of technology during and after conflict in Cambodia, and an examination of the gendered consequences of agricultural biotechnology in Kenya.

Gender and technology: an overview

Historically, technology was categorized as opposing or controlling the "natural" world. Technology was associated with modernization and mechanization, and moving away from "nature" and traditional livelihood strategies. Separating "technology" from "nature" established a representational trend that situated the use of technology as a sign of development and subsequently associated "nature" and older technological processes with the need for technologically driven economic development. Many of these technologies were gendered, whereby technological skills were associated with men and masculinity, and nature or the natural world were coupled with women and femininity (Rose 1993). Feminist scholars have argued that the spilts between nature/technology and developed/developing worlds remain false binaries (Trauger 2004). Just as technology is a product of human invention, contemporary conceptualizations of "nature" have been "invented." Defining what is "natural" and "technological" depends on social and political perspectives rather than an inherent or clearly demarcated division between the two. Separating technology and nature into a binary framework is a method for claiming progress or lack thereof and subsequently used to support the "need" for development. While technology is often lauded as an example of human innovation and progress, many contemporary technologies (i.e., factory automation, information technology, transportation) depend upon fossil fuels that are inextricably linked to global climate change and environmental degradation. Thus, technology has provided new opportunities as well as environmental problems and adding additional burdens and difficulties for people in various societies. The development and use of technology have shaped and are shaped by intersectional gender roles/relations.

Ester Boserup (1910–99) was a late 20th-century feminist scholar who identified how modern technologies increased male participation in the various economic markets and technologically driven economies, which subsequently added to women's work burdens within their families and through casual and unpaid labor. More recently, the increase of women within economic development programs included recruiting women into modernization and mechanization projects to accelerate economic growth. In other cases, technology has led to job losses for both men and women. For example, the 1986 and 1994 UN global surveys on women in development found that technological modernization failed to improve women's productivity because it bypassed them and pushed them out of jobs in favor of new technologies such as automation (Jain 2005).

In other cases, technology has been used to push back against social structures that disadvantage certain groups based on gendered, classist, caste-based and racist assumptions. For example, scientific knowledge was used to challenge the caste system in India. B.R. Ambedkar, a **dalit** reformer, challenged the ideology of the **caste system** by using concepts of scientific knowledge and "objective truth." Ambedkar identified the caste system as an invention of Hindu mythology, and he "was convinced that only a secularization of consciousness through a rigorous application of reason and scientific temper could bring about a transformative change" (Nanda 2004, 217). Conversely, some feminist critiques of development technologies highlight the ways in which patriarchal structures remain fundamental to modern technologies (Benería 2003, Benería, Berik and Floro 2016). These critiques show that men dominate various technological sectors, particularly higher-paid jobs requiring specialized skills. Lower-skill and assembly-line work in factories has become increasingly feminized in certain places, meaning women remain sought after as workers for this type of labor. The feminization of some forms of employment is often attributed to liberal feminist ideologies that pursue women's economic participation as a method of emancipation from patriarchy and a path toward social or political liberation. Increased access to technology, education and paid labor have improved the lives of some women, while other examples of feminized labor are highly exploitative, demanding long hours of employment, restrictions on mobility and family planning, and providing low and irregular wages (Cornwall, Harrison and Whitehead 2007).

Technological opportunities for improving roads, manufacturing and other forms of production have, in general, been identified as foundational for enhancing a country's long-term economic growth (Sheppard et al. 2009). Early forms of international development sought the transfer of skills from technologically rich countries to countries with minimal mechanization or technological modernization. Colonizers viewed colonized places and people as "traditional," "backward" and "non-modern" partly based on lack of technological modernization or mechanization. Thus, technology became a method through which colonizers identified themselves as "superior" and colonized areas as "inferior." Colonizing powers situated binary representations of places as either modern or traditional. This dichotomous assessment of places influenced early development efforts that sought to "assist" places through large-scale mechanization and "modern" infrastructure projects. Dam construction for the purposes of hydroelectric power was one area of economic and geopolitical significance to the United States. The United States sought to produce and perpetuate the production of large-scale dams in Asia, Africa and Latin America from the mid-20th century up until the early 1980s (Sneddon 2015). These and other development technologies prior to the mid-1970s were mostly focused on large-scale infrastructure projects, and male-dominated.

Male dominance in the engineering profession subsequently focused on infrastructure projects that addressed the needs of men without the inclusion of women and girls. The introduction of new technologies has in some cases increased men's participation in tasks and skills that had previously been done exclusively by women (Hart 1992, Naylor 1994, Davis 2005). While programs to address women and gender, such as WID, WAD and GAD (see Chapter 2), began in the 1970s, gender pluralism within development technologies remains minimal. However, women have been increasingly part of the low-wage workforce in mechanized factories, particularly in the garment industry. For example, in Bangladesh there are more than 4,000 garment factories employing four million workers, 80 percent of whom are women (Caeson 2015). While in some cases factory work provides women with more economic opportunity and autonomy, research has also shown that it can mirror prevailing gender imbalances and inequalities (Mies 2014); for example, when male managers operate within patriarchal frameworks that expect female workers to be docile, obedient and defer to men (Wright 2006).

Labor and technology: gendered development meets globalization

When examining technology – as a part of economic development practices – it is imperative to also examine geopolitics in order to understand how politics shapes how technology is conceptualized, practiced and praised. It is similarly important to identify the suitability of technology for women and men in different communities, sociocultural contexts and life stages (Radcliffe 2015, 289). In some places, men and women have different access to technological resources and decision-making abilities (Skutsch and Clancy 2006). While efforts have been made to incorporate gender-sensitive approaches to technology transfer through development projects, these approaches remain minimal when compared with other sectors such as health, education and employment.

Globalization as an extension of economic development practices includes the use of foreign direct investment (FDI) and the growth of transnational corporations (TNCs). Both FDI and TNCs seek low-waged labor in countries with new or emerging forms of industrialization, a surplus labor force, and lack of or restriction on labor unions. Research on TNCs identifies both positive and negative effects for men and women globally. In some instances, manufacturing jobs provide new forms of income and opportunity. However, low-wage and unregulated labor within many manufacturing companies includes physical abuse of employees, the use of child and slave labor, unventilated, toxic working conditions and long hours for minimal pay (Sparke 2013). Many companies that employ female laborers for assembly-line manufacturing jobs envision certain jobs as gendered (i.e., more appropriate for men in some cases and women in other cases). The gendering of work includes: 1) paying women less than men because female salaries are viewed as supplemental rather than primary income within their households; 2) hiring women on cyclical 2–4-year contracts; and 3) viewing women as docile workers rather than leaders or managers.

Research on manufacturing jobs underscores gender divisions of labor in many sectors and the increased use of female labor based on the expectations that women will be more docile and obedient than men. For example, in the **maquiladoras** (manufacturing operations in Mexico's "free trade" zones where material and equipment are free from duty and tariffs), many companies hire young unmarried women

FOCUS: CHINA

In China, the development of a capitalist economy along with a socialist state has created a system of migratory laborers known as *mingong* (peasant workers). These workers are bound to systems of dormitory labor in developing and industrialized spaces within China. This system of temporary/migratory labor in China is legitimized by the state which "provides population and labor control that favors global and private capital" (Ngai 2005, 5). Gender relations, while changing, remain patriarchal in many rural spaces, particularly in terms of gender divisions of labor, educational opportunities and marriage age (Ngai 2005). In many cases, a young (usually in her 20s) woman's family decides whether or not to allow her to work in urban manufacturing centers. Short-term waged labor is fairly typical for rural women between the ages of 18 and 25. Factory work has therefore been constructed through gendered expectations of the female laborer. For example, "as a girl in the process of becoming a woman, one should behave as the culture required: submissive, obedient, industrious, tender and so on. . . . Maleness was posited as a degraded opposite in warning to workers 'you should not act like a boy, a boy is lazy and troublesome, careless and rough. Otherwise, you can't marry yourself out'" (Ngai 2005, 144). Therefore, as this quote suggests, both men and women were categorized based on socially constructed gender roles in an effort to ensure a gendered ordering of industrialized labor and to remind women that their ultimate goal was to eventually marry and return to their rural villages.

on two-year time cycles in order to ensure a revolving door of employees who are fired or laid off before they can organize a union or demand better conditions, pay and hours (Wright 2006). In these factories, women who are mothers are seen as less desirable employees. Employers assume that mothers will take significant time off from work to care for their children, while the same standards are not applied to male workers who are fathers. Thus, women's domestic responsibility, both actual and perceived, remains an expected part of their gendered paid and unpaid labor. Fathers are not expected to take time away from work because it is assumed that their children will be minded by women (i.e., mothers/grandmothers). These assumptions further reinforce the belief that mothers cannot or should not work the same hours as women without children or men; therefore, being a mother continues to present a barrier for women seeking certain kinds of employment. For single mothers, these barriers are particularly difficult as they are both caregivers and the only economic producers for their families. Additionally, backlashes against the increased number of women employed in factories and moving in public spaces at various times of the day and evening (due to shift work) have been factors in the growing epidemic of violence against women working in the maquiladoras (Martin and Carvajal 2015).

Historically, factory labor was male-dominated. However, due to changing economic patterns, increasing numbers of women are working as laborers in factories throughout the globe. Despite and in association with these changes, women continue to occupy a secondary employment position (in terms of power and income) to men. In many cases, existing patriarchal structures and hierarchies are purposely reproduced within factories in order to discipline female employees.

Information and communication technologies

Information and communication technologies (ICTs) similar to other forms of technology – such as factory automation – are assumed by development organizations as a pathway toward empowering for women (Ng and Mitter 2005). While women's access to ICTs can provide economic opportunities, ICTs can

Figure 8.1 Maquiladora factory

Source: by Guldhammer [CC BY-SA 3.0 (https://creativecommons.org/licenses/by-sa/3.0)], from Wikimedia Commons

also further exacerbate rather than alleviate gender, race and economic inequalities (Ng and Mitter 2005). Despite increases in accessibility to ICTs globally, a digital divide persists with the majority of the world's population continuing to experience limited to no access to ICTs, which excludes rather than includes them as part of the Internet-networked world. Women's and men's access to ICTs has increased dramatically in many countries. Attempts are being made to change the gender imbalances associated with ICTs, which often require addressing existing gender inequalities within a given society. For example, despite the University of Khartoum (Sudan) declaring equal access to computers for all students, social views that women should be less interested in ICTs than men resulted in men having more experience with these technologies (Buskens and Webb 2014). In India, political approaches to improving women's rights include improving women's access to the Internet and other ICTs (Gurumurthy, Chami and Thomas 2016). Meanwhile, a study in Ghana, Kenya and Kerala, India, revealed that women's access to technology continues to lag behind that of their male peers (Miller, Duque and Shrum 2012).

Patel (2010) examined call centers in India to determine the spatial and social difficulties of communication technologies, gender and work. Her research shows that while call centers provided economic opportunities for women, working outside the home has been much more complex for these women. Beginning in the 1990s, Fortune 500 companies in the United States began moving customer service and call center jobs to India as part of the wave of globalization known as business process outsourcing (BPO). India offered a low-wage labor force with English language skills (a legacy of British colonialism in India). Due to the extensive time differences between the United States and India, most of the call center jobs in India required overnight shift hours (i.e., 10 pm to 6 am, or 8 pm to 4 am). This created physical, temporal and spatial difficulties for workers who needed to travel to work at times generally designated for being at home. Traversing public space in the evening is generally viewed as inappropriate for women and therefore many companies provide transportation as a method for recruiting female employees.

Despite these efforts, many Indian women who work at night are subject to physical and sexual harassment, abuse and threats from strangers and sometimes male supervisors. Call centers reveal shifting relationships between gender, technology, work, space and time. For example, spaces are dynamic and can change over time, including over the course of 24 hours. A safe space during the day can become insecure at night. Designating certain spaces as insecure or inappropriate for women is also a way to control or reduce women's mobility. Therefore, when women are assaulted in spaces designated as insecure, victim blaming often occurs, rather than focusing on the perpetrators' actions or the need to ensure security in these spaces at all times. Similar to maquiladoras, women must prove their legitimate right to traverse public spaces (especially at night) rather than delegitimizing the use of violence by men in public space. Due to national and international activism, women are claiming their right to traverse public space and work outside the home in these and other contexts. Call centers also expect their employees to perform what Mirchandani (2012) identifies as "authenticity work." Authenticity work requires employees to interact with callers in ways that are both similar to and different from them. Thus, workers are continually dealing with how to "act" more Western to meet the needs and expectations of callers, while still maintaining their "authentic" Indian identity (Mirchandani 2012).

In addition to call centers, other types of ICT jobs in India are gendered and require workers' spatial mobility. For example, "body shopping" is a common practice among ICT professionals in India. "Body shopping" is a colloquial term used by ICT professionals to describe the practice of contract work, where one company will "lend out" an ICT worker to another company to complete specific tasks or projects (Biao 2007). Men dominate "body shopping" programs. Few women participate in these schemes due to concerns about gender-based violence against female workers from their male counterparts (Biao 2007). However, wives have been integral to the process of setting up "body shops" and providing steady income when male workers are between projects. This type of technological labor migration has been constructed and sustained through gender relations. For example, married women's paid labor helps to provide steady and sustainable work to offset the highly unstable forms of international "body shopping" for ICT workers. Cultural value has been attributed to this and other forms of ICT work. For example, higher dowries are often demanded of families seeking to marry their daughters to an ICT worker.

The above examples show that existing gender divisions of labor, space and opportunity may be transformed by technology in some cases, and in others exacerbate inequalities or create new forms of spatial, social and situational gendered divisions. Changing gendered economic patterns has increased the number of women entering the workforce globally. In some places, this has led to significant backlashes, including the use of violence. In response to this violence, many women are identifying, claiming and demanding their legitimate right to move through public spaces at different times without hesitation, harassment or harm. ICTs identify global inequalities associated with waged labor, while biological technologies highlight the extensive divisions between wealthy and poor individuals along with racial inequalities that have become integral to global systems of exchange.

Biotechnologies: gender, race, class and reproduction

Advances in reproductive technology have included an increased use of birth surrogates. In recent years, gestational surrogacy has followed the development of global assembly lines and the new international division of labor. These industries capitalize on the reproductive capacities of poor women. In March of 2017 the state of India banned the use of commercial surrogates for all childless couples living in other countries, and the surrogacy industry banned gay couples from participating in 2012. The industry has also been rocked by scandal, legal battles and fraud, including failure to deliver babies to expectant couples and deaths of infants. Bioethicists welcome the ban, arguing that profit should not be made from pregnancy and reproduction, and questioning the way it puts a monetary value on human life. Justification for the use of poor women follows arguments made for employing women in other kinds of poorly regulated and exploitative work, i.e., that it is lucrative for married women with children who otherwise may not be able to work due to lack of training, education or childcare.

Reproductive technologies (as discussed in Chapter 7) provide women and men with various options for family planning. In this section, we explicate the ways in which these technologies operate as a political tool to control, empower, liberate or marginalize different individuals and groups. Technology alone cannot improve economic conditions, and in several cases it exacerbates existing economic divisions. For example, the use of surrogates for biological reproduction exemplifies global economic divisions. International surrogacy has grown more common, where relatively wealthy women who are unable (or choose not) to carry a fetus to term offer to pay economically disadvantaged women to act as surrogates. The fetus is the biological product of the wealthy woman or couple, which is artificially placed into the body of a woman who carries the child to term. Reproductive technologies assist with this form of womb outsourcing. The female surrogate bears the physical body work of carrying a fetus to term and giving birth.

Economically, the brokers who set up these transactions and the reproductive centers that conduct the procedures receive the majority of the money paid by the individual or couple requesting this service. The female surrogate receives the smallest percentage of the overall fee paid. There are minimal regulations on international surrogacy, as laws vary significantly by country. Internationally, regulation has not yet caught up to innovation in surrogacy practices. In many cases there is "mutually advantageous exploitation whereby prospective parents or surrogacy brokers unfairly benefit from the surrogacy transaction at the expense of the surrogate mothers" (Humbyrd 2009, 114). In order to address reproductive justice in the surrogacy marketplace, it is imperative to analyze existing trends through an "intersectional analysis of race, gender and class" (Mohapatra 2012, 5). For example, surrogates are predominantly poor women from marginalized racial groups, and in India from lower castes. Brokers are from a higher socioeconomic class and country and do not experience as much racial or caste discrimination. They solicit women in situations of economic privation with the promise of medical care and a large payment (relative to what the women could earn within the same period of time). Research on surrogates in India highlights that surrogates are themselves produced as "cheap, docile, selfless and nurturing" through the structures and trainings in fertility clinics and surrogacy hostels (Pande 2010, 970). Thus, a surrogate mother is subjected to similar disciplinary processes used for training factory workers (Pande 2010). Surrogates remain intersectionally vulnerable to brokers and clinics that view their bodies as sites of capital accumulation. However, surrogates are not without agency and in some cases use their experiences and parlay them into other forms of employment or collective organizing. As Pande's (2010, 971) research identifies:

> The hostels constitute a gendered place, one that generates emotional links and sisterhood among the women. This intensive contact allows the surrogates to share information and grievances with one another and to sometimes come up with strategies for future employment and even acts of collective resistance.

In addition to reproductive surrogacy, a shortage of organ donors and global inequalities have led to the development of the international organ trade. The General Agreement of Trade in Service (GATS) allows governments to meet their national health objectives through trade in health services, which some countries implement as a form of economic development. Potential donor recipients travel to another country in order to receive an organ donation through commercial suppliers. This practice has become so common and exploitative to poor individuals donating their organs that it prompted the World Health Assembly to adopt a resolution in 2004 for its member states to "take measures to protect the poorest and vulnerable groups from 'transplant tourism' and the sale of tissues and organs" (WHO 2004, 2). Regulations have been instituted in countries such as India, which was a major organ-exporting country; however, underground trafficking in organs continues. China is another country with a significant number of kidney and liver transplants, while foreign recipients have decreased in recent years. Kidneys are reportedly sold in several countries, including Bolivia, Brazil, Iraq, Israel, the Republic of Moldova, Peru and Turkey (Scheper-Hughes 2009). Similar to organ donation, stem cell development companies "operate between and betwixt national regulatory environments" (Dixon 2015, 63). Thus, a global trade in stem cells occurs in an effort to allow individuals to avoid regulatory procedures in one country by traveling to another. "Stem cell treatment so often requires an intricate array of mobilities, with materials crossing a series of sub- and inter-national borders" (Dixon 2015, 64). Globally, women are more likely than men to donate their organs, while men are more likely than women to accept transplant surgery (Puoti et al. 2016). There are several theories as to why these gender disparities exist, from sociocultural norms to differential medical needs for men and women. However, due to the extensive lack of regulation of global organ trade, more research is needed to understand the gendering of organ donation and trade in other biomaterials.

Each of the above examples points to international trade regimes that incorporate poor and racially or socially marginalized bodies into the paid service of corporeal biotechnologies that extend the lives of wealthy individuals and shorten the lives of mostly poor donors. These trade regimes do not take into consideration the long-term health effects on individuals who sell or rent parts of their bodies as a method of ensuring economic opportunity. The following two case studies (Cambodia and Kenya) extend the discussion of biotechnologies further. The Cambodian case study examines the ways in which gender, conceptions of beauty and international "assistance" intersect with technologically advanced prosthetics through the Miss Landmine pageant. The second case study focuses on biotechnology in Kenya through the development of genetically modified seed technologies.

Case studies: Cambodia and Kenya

These two case studies analyze different forms of biotechnologies: prosthetic and agricultural. Advances in prosthetic biotechnologies have increased in recent decades, particularly with increased funding from some countries. The United States has increased its funding of prosthetic science and technology to assist the large numbers of servicemen and women who have returned from the wars in Afghanistan and Iraq with amputated limbs. The case study on Cambodia focuses on a small organization that ran a beauty pageant for female amputees in an effort to call attention to the high number of land mine injuries and deaths there. The Kenya example focuses on the use, misuse and government control of agricultural biotechnologies. Agricultural technology such as genetically modified plants and pesticides has spread throughout the globe through various economic development programs. The Kenya case study exposes several problems associated with these technologies.

Cambodia

Racial, gender and economic inequalities are exacerbated during war and its aftermath. Conflict development (see Chapter 9) relies on the progress of military technologies during active conflict and afterwards to maintain the bodies of soldiers and citizens with war-related injuries. War-related damage includes the disruption of infrastructure and access to technologies such as electricity, running water, proper sanitation and the Internet. War technologies such as land mines and cluster bombs continue to maim and injure civilian bodies well after the conflict has ended. Governments with limited access to economic resources may receive assistance from international mine removal organizations, but demining remains a slow and methodical process. Individuals who survive a land mine explosion are often left with serious injuries, including amputated legs or arms, traumatic head injuries and blindness. Biotechnologies associated with bodily recovery and prosthetics for amputees have improved in the past two decades. The US government has provided significant funding to improve prosthetics due to the large number of soldiers returning from war with amputated limbs. Bionic and mechanical prosthetics today provide new forms of mobility for amputees. However, these technologies remain exorbitantly expensive and therefore out of reach for many amputees, especially those living in poverty in the United States and throughout the globe.

Cambodia continues to experience the consequences of land mines and cluster bombs from three decades of war during the Khmer Rouge, Heng Samrin and Hun Sen regimes (Tyner 2009, 2017). Active land mines have killed and injured tens of thousands of people in Cambodia. Many of the injured survivors of land mines are missing one or more limbs. The majority of land mine victims are men and boys. In Cambodia 87 percent of victims are males over the age of 15. Afghanistan is another country with four decades of conflict and an extensive number of land mines throughout the country. Seventy-three per cent of land mine victims in Afghanistan are males between the ages of 16 and 50, and of the 73 percent, 20 percent are male children. The overwhelming majority of these victims do not have access to high-tech prosthetics and bionic limbs. These biotechnologies exemplify significant technological advancements and highlight corporeal divisions between wealth and poverty. In spaces of protracted conflict, there are higher rates of amputees and endemic poverty than in spaces such as the United States, which is continually engaged in political conflict but not on its own soil.

In an effort to call attention to the land mine crisis in Cambodia, the all-female Miss Landmine beauty pageant was launched by a male Norwegian activist. While there are far fewer female than male land mine victims, this pageant exemplified the ways in which women's bodies are spaces upon which social, political and economic ideologies are advertised. In the contestant catalogue for the Cambodian pageant, images of each woman dressed up in an American Apparel outfit were shown with text that identified the cost of the dress, shoes and accessories, and by contrast the cost of the land mine munitions that caused her injury. The contestant information further included her home location and that of the land mine explosion that injured her body, as well as the land mine's country of origin/manufacture. In this catalogue, the contestants' bodies provided visually striking methods for illustrating and detailing the costs and geographies associated with war technologies (Fluri 2018).

This pageant used performance, image and film as methods for "challenging" conventional beauty expectations for pageant competition and the viewer's ability to see the disabled and amputated war body as beautiful. However, the conceptualization of the pageant did not challenge the ways in which women's bodies are objectified through pageantry or economic inequalities that prevent the contestants from receiving technologically advanced prosthetics. The winner of the Miss Landmine pageant received a biotechnology that remains economically unavailable to her and the other contestants, a titanium leg worth approximately $15,000. The Miss Landmine website positioned "beauty" under a human rights umbrella, as suggested by the tagline: "Miss Landmine: everyone has the right to be beautiful." However, there was no discussion of everyone's right to biotechnologies that would improve their mobility and quality of life. Rather, the pageant reduced the needs of their injured bodies to a competition that provided just one woman, the winner, with titanium prosthetic biotechnology. In this example, similar to others, technology highlights existing inequalities rather than addressing or "fixing" them. The next case study focuses on agricultural biotechnologies as an example of technology transfer through development and the limits of these technologies to address the problems they intend to solve.

Kenya

Agricultural biotechnology remains a controversial method of technology transfer. Scholars of development focused on food insecurity and famine extensively debate the effectiveness of agricultural biotechnology to create food security in developing countries (Harsh and Smith 2007). Development organizations and financial support institutions are divided on whether or not biotechnologies are beneficial: "NGOs such as

Action Aid argue that there are potentially no material benefits, only risks being gained from agricultural biotechnologies" (Harsh and Smith, 2007, 251). However, other donor countries and organizations have supported the development and implementation of biotechnologies in various locations. In Kenya, the growth of agricultural biotechnology has been donor-funded and donor-led (Harsh and Smith 2007). The history of expecting "technological fixes" to improve economic conditions in developing countries "makes top-down approaches a persistent feature of agricultural development in Kenya" (Odame 2002, 2749).

Beginning in 1991, the biotech company Monsanto and USAID both sponsored biotechnology in Kenya for the production of virus-resistant sweet potatoes. Despite initial failures this project continued and other crop varieties were tested, such as maize, cassava and cotton. By 2004, a New Partnership for African Development (NEPAD) began a Nairobi-based program to focus on biotechnological research and development (Harsh and Smith 2007, 253). NEPAD highlighted the role of women (along with men) as necessary for decision making and environmental management (Kameri-Mbote 2007). Additional funding flooded into Kenya (often at odds with NEPAD), including from large donor organizations such as the Rockefeller Foundation and the World Bank. These international agencies were able to accelerate research, development and implementation of biotechnologies due to Kenya's developed infrastructure and *lack* of government oversight. Therefore, many organizations and corporations produce biotechnologies within a "'legislative vacuum,' which leads to a lack of transparency and accountability in decisions about biotechnology – a lack of good governance" (Harsh and Smith 2007, 253). Without legal oversight for international donors and research institutes, they are not held accountable to farmers or the public, even when their projects violate biosafety rules (Harsh 2005).

The organizations implementing chemicals for pest management did not provide adequate training or information to famers. This was particularly evident for the Gwitheria Women's Group. These women experienced suspicion from husbands about agricultural pesticides because they were unlabeled. "Such problems demonstrate not only the need for better packaging and labeling, but also the necessity to understand the social and gender issues that impinge on technology diffusion" (Odame 2002, 2752). Additionally, farmers were concerned that chemicals posed health risks, particularly to young children and farm animals. For example, in Kenya, women's exposure to chemicals in the agricultural sector has been linked to miscarriages (Kameri-Mbote 2007). These chemicals raised the farmers' costs because of the higher labor demand and much slower rate for planting (Odame 2002). The creation of "subjects" in need of biotechnology in Kenya (and elsewhere) also grossly overlooks the way in which small farmers, who are disproportionately women, cannot use (often because they cannot afford to use) "improved" seeds (Parker 2017). Additionally, biotechnologies in many African countries are meant to enroll farmers in capitalist development practices and patterns (i.e., large-scale, chemical-intensive farming).

Biotechnologies, similar to other forms of technology transfer, do not take into consideration sociocultural practices and intersectional gender roles and relations. Some have argued that the potential for biotechnology "to increase agricultural production in Africa lies in its 'packaged technology in the seed,' which ensures technology benefits without changing local cultural practices" (Stone 2010, 389). However, other researchers contend that the self-contained seed is not as revolutionary as suggested because farmers use more nuanced methods for selecting seeds, and the effects of genetically modified seeds are integral to a variety of farming institutions (Stone 2010). Biotechnology is designed to increase agricultural production for sale and export, which undermines seed varieties (and continued seed variability) and in some cases marginalizes women from performing their expected responsibilities associated with both food production and biological and social reproduction (Kameri-Mbote 2007). The extensive and complex environmental knowledge systems women have developed are often overlooked. Women who are members of marginalized communities also suffer from "intersectional discrimination," a double experience of ostracism as women and minorities (Kameri-Mbote

Figure 8.2 Biotechnology laboratory in Nairobi, Kenya

Source: courtesy of Lowery Parker

2007, 41). Additionally, the gendered impact on communities includes specific vulnerabilities for women when they are pregnant and nursing in addition to the social marginalization of women from the means of production.

The use of biotechnology in Kenya continues, while the Kenyan government has increased its level of oversight and monitoring of these developments, e.g., the enactment of the Biosafety Act in 2009 and launch of the National Biosafety Authority in 2010, which published new biosafety regulations 2011–2012. Kenya's 2012 food safety law significantly limited the development and use of agricultural biotechnology, including halting the open field trials for genetically modified crops, such as Bt corn (Kamau 2016). The "technological fix" expected of biotechnologies such as chemical fertilizers, pesticides and genetically modified crops has not lived up to the "hype" that they would increase crop yields for sale and export. Several concerns remain about the cost–benefit analysis of biotechnologies, particularly when chemicals and modified genetic varieties cause lasting detrimental effects to people (particularly women's reproductive bodies), land/soil, water and animals.

Emancipatory technologies?

Biomass fuels such as dung cakes, wood and coal continue to be used as sources for cooking in many parts of the world. Technological innovations have converted organic materials into biogas, which provides more efficient and cleaner burning fuel for household use. Biogas has been produced in Asia (mostly India and China) since the late 19th century. It is created from the breakdown of organic matter in the absence of oxygen and can be produced from raw materials such as manure, organic waste, plants, food waste

Figure 8.3 Woman working over a coal stove

and sewage. It can be used for electricity production, cooking, space heating and water heating, and may be compressed for use in vehicles as an alternative to gasoline. The use of biogas for cooking has significantly improved conditions in many places that previously used conventional forms of biomass such as wood, dung and coal. In rural communities in China where women are responsible for cooking household meals, biogas digesters for cooking fuel have had a positive impact on their health, lessened the intensity of their work and improved living standards. The use of biogas and solar energy cookers has reduced the time spent and frequency of collecting firewood and reduced the risk of indoor smoke pollution, which causes respiratory disease (Ding et al. 2014). While this has had a positive impact on women, the introduction of new technologies such as biogas or other energy producing machines does not necessarily change power divisions, gender divisions of labor, emancipate women or reduce poverty.

Biogas provides alternative energy options that have the potential to meet energy needs through the use of renewable rather than fossil fuels. Alternative energy technologies are often marginalized or sidelined by national and multinational corporations investing in and profiting from the consumption of fossil fuels. The highly profitable business of nonrenewable resources views alternative fuels, such as biogas, as a threat to their market share. Biogas has already had a positive impact on several communities, particularly for women due to existing gendered divisions of labor (i.e., women cooking and ingesting less noxious fumes from biogas as compared to burning other organic materials). Therefore, while biogas may indeed be a

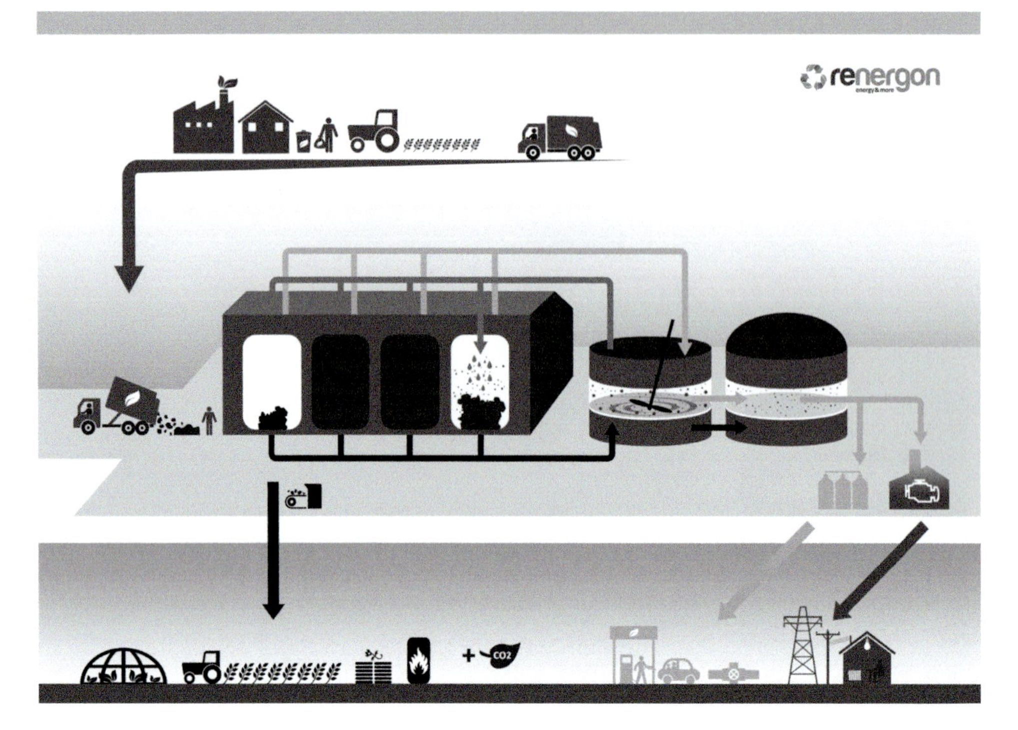

Figure 8.4 Biogas diagram

Source: by Renergon International AG [CC BY-SA 3.0 (https://creativecommons.org/licenses/by-sa/3.0)], from Wikimedia Commons

viable and preferable alternative to biomass and fossil fuels, political and economic structures often operate as a barrier rather than conduit to increasing the use, development and effectiveness of biogas. Social and economic marginalization can keep individuals (based on the intersectional categorizations such as gender, race, class, sexuality, dis/ability) from being able to access these technologies.

Technology offers much promise for various communities throughout the globe. In addition to progress, technology has created more challenges and difficulties for many communities. The transfer and use of various technologies are intersectionally gendered and may create or exacerbate existing social and identity-based inequalities. Technology does not always "fix" identified problems, and in many examples reinforces social and spatial divisions or enrolls poor, sociopolitically marginalized and disenfranchised persons into forms of labor that benefit individuals with more wealth and resources. Because technology remains embedded within capitalist modes of production and consumption, it often exemplifies local and global divisions among gender, racial and class groups.

Recommended reading

Discounted life: the price of global surrogacy in India, Sharmila Rudrappa; *Women and the machine*, Julie Wosk

Recommended viewing

Made in India

Questions for discussion

Identify different ways in which technology exemplifies gender, race and class divisions within your society. What do you think can be done to mitigate these divisions? What would a noncapitalist approach to technology look like? What would need to happen to increase gender pluralism within various technological sectors? In what ways does technology improve your life? In what ways does technology make your life more difficult? Can you imagine going a year, month, week, day or hour without the use of technology? Identify all the ways in which you use technology in a given day.

References

Benería, L. (2003). *Gender, development and globalization: economics as if all people mattered.* New York and London: Routledge.

Benería, L., Berik, G., & Floro, M.S. (2016). *Gender, development and globalization: economics as if all people mattered.* New York and London: Routledge.

Biao, X. (2007). *Global "body shopping": an Indian labor system in the information technology industry.* Princeton, NJ: Princeton University Press.

Buskens, I., & Webb, A. (Eds.) (2014). *ICTs and gender equality: transformation through research?* London: Zed Books.

Caeson, B.S. (2015). Our voices, our safety: Bangladeshi workers speak out. *International Labor Rights Forum*, December: 1–105.

Cornwall, A., Harrison, E., & Whitehead, A. (Eds.) (2007). *Feminisms in development: contradictions, contestations and challenges.* London and New York: Zed Books.

Davis, D.K. (2005). "A space of her own: women, work, and desire in an Afghan nomad community." In G.-W. Falah, & C. Nagel (Eds) *Geographies of Muslim women: gender, religion, and space*, New York and London: Guilford Press (pp. 68–90).

Ding, W., Wang, L., Chen, B., Xu, L., & Li, H. (2014). Impacts of renewable energy on gender in rural communities of north-west China. *Renewable Energy*, 69, 180–89.

Dixon, D.P. (2015). *Feminist geopolitics: material states.* Surrey: Ashgate Publishing Company.

Fluri, J.L. (2018). "Feminist political geography and geopolitics." In A.M. Oberhauser, J.L. Fluri, R. Whitson, & S. Mollett (Eds) *Feminist spaces: gender and geography in a global context*, London: Routledge (pp. 131–54).

Gurumurthy, A., Chami, N., & Thomas, S. (2016). Unpacking digital India: a feminist commentary on policy agendas in the digital moment. *Journal of Information Policy*, 6, 371.

Harsh, M. (2005). Formal and information governance of agricultural biotechnology in Kenya. *Journal of International Development*, 17(5), 661–77.

Harsh, M., & Smith, J. (2007). Technology, governance and place: situating biotechnology in Kenya. *Science and Public Policy*, 34(4), 251–60.

Hart, G. (1992). Household production reconsidered: gender, labor conflict, and technological change in Malaysia's Muda region. *World Development*, 20(6), 809–23.

Humbyrd, C. (2009). Fair trade international surrogacy. *Developing World Bioethics*, 9(3), 111–18.

Jain, D. (2005). *Women, development, and the UN: a sixty-year quest for equality and justice.* Bloomington, IN: Indiana University Press.

Kamau, C.N. (2016). *Kenya's agricultural biotechnology report.* Global Agricultural Information Network. USDA Foreign Agricultural Service. Washington, DC.

Kameri-Mbote, P. (2007). Access, control and ownership: women and sustainable environmental management in Africa. *Agenda: Empowering Women for Gender Equity: Two Decades of African Feminist Publishing*, 72, 36–46.

Martin, P., & N. Carvajal (2015). Feminicide as "act" and "process": a geography of gendered violence in Oaxaca. *Gender, Place & Culture*, 23(7), 989–1002.

Mies, M. (2014). *Patriarchy and accumulation on the world scale: women in the international.* 3rd edition. London and New York: Zed Books.

Miller, B.P., Duque, R., & Shrum, W. (2012). Gender, ICTs, and productivity in low-income countries: panel study. *Science, Technology, & Human Values*, 37(1), 30–63.

Mirchandani, K. (2012). *Phone clones: authenticity work in the transnational service economy.* Ithaca, NY: Cornell University Press.

Mohapatra, S. (2012). Achieving reproductive justice in the international surrogacy market. *Annals of Health Law*, 21, 1–7.

Nanda, M. (2004). "Do the marginalized valorize the margins? Exploring the dangers of difference." In K. Saunders (Ed.) *Feminist post-development thought: rethinking modernity, post-colonialism, and representation*, London and New York: Zed Books (pp. 212–24).

Naylor, R. (1994). Culture and agriculture: employment practices affecting women in Java's rice economy. *Economic Development and Cultural Change*, 42(3), 509–35.

Ng, C., & Mitter, W. (Eds) (2005). *Gender and the digital economy: perspectives from the developing world.* New Delhi: Sage Publications India.

Ngai, P. (2005). *Made in China: women factory workers in a global workplace.* Durham, NC: Duke University Press.

Odame, H. (2002). Smallholder access to biotechnology: case of Rhizobium Inocula in Kenya. *Economic and Political Weekly*, 37(27), 2748–55.

Pande, A. (2010). Commercial surrogacy in India: manufacturing a perfect mother-worker. *Signs: Journal of Women in Culture and Society*, 35(4), 969–92.

Parker, L. (2017). Modified states: sovereignty and the ethics of crop biotechnology in Kenya. Unpublished PhD thesis. University of Georgia.

Patel, R. (2010). *Working the night shift: women in India's call center industry*. Stanford, CA: Stanford University Press.

Puoti, F., Ricci, A., Nanni-Costa, A., Ricciardi, W., Malorni, W., & Ortona, E. (2016). Organ transplantation and gender differences: a paradigmatic example of intertwining between biological and sociocultural determinants. *Biology and Sex Differences*, 7(35), 1–5.

Radcliffe, S. (2015). *Dilemmas of difference: indigenous women and the limits of postcolonial development policy*. Durham, NC: Duke University Press.

Rose, G. (1993). *Feminism and geography: the limits of geographical knowledge*. Minneapolis, MN: University of Minnesota Press.

Scheper-Hughes, N. (2009). Organs without borders. *Foreign Policy*. https://foreignpolicy.com/2009/10/21/organs-without-borders/. Accessed 2/17/2019.

Sheppard, E., Porter, P.W., Faust, D.R., & Nagar, R. (2009). *A world of difference: encountering and contesting development*. New York and London: Guilford Press.

Skutsch, M., & Clancy, J. (2006). "Unraveling relationships in the energy-power-gender nexus." In J. Byrne, N. Toly, & L. Glover (Eds) *Transforming power: energy, environment, and society in conflict*, London: Transaction Publishers (pp. 61–92).

Sneddon, C. (2015). *Concrete revolution: large dams, cold war geopolitics, and the US bureau of reclamation*. Chicago, IL: University of Chicago Press.

Sparke, M. (2013). *Introducing globalization: ties, tensions, and uneven integration*. West Sussex: Wiley-Blackwell.

Stone, G.D. (2010). The anthropology of genetically modified crops. *Annual Review of Anthropology*, 39(1), 381–400. https://doi.org/10.1146/annurev.anthro.012809.105058.

Trauger, A. (2004). Beyond the nature/culture divide: corporeality, hybridity and feminist geographies of the environment. *WGST, Geography and Gender Reconsidered*, August, 21–34.

Tyner, J. (2009). *War, violence, and population: making the body count*. New York, NY: Guilford Press.

Tyner, J. (2017). *The business of war: workers, warriors and hostages in occupied Iraq*. London: Routledge.

WHO (2004). Human organ and tissue transplantation. Agenda item 12.14. www.who.int/transplantation/en/A57_R18-en.pdf. Accessed 2/15/2019.

Wright, M.W. (2006). *Disposable women and other myths of global capitalism*. New York: Routledge.

9 Disaster assistance and development

Introduction

Providing assistance to the citizens of a country in the aftermath of a disaster is a form of **soft power**. This form of power can and often does resonate through various forms of superiority (e.g., racial, Western, moral) (Grewal 2014). This form of assistance is often geopolitically motivated to secure strategic alliances or control in a geographic area of interest. As discussed in Chapter 3, the business of international assistance development is also stymied by competing and contradictory interests among governmental and non-governmental organizations (NGOs). The divergent geopolitical and institutional requirements of various organizations can "promote destructive competition among well-meaning transnational actors" (Cooley and Ron 2002, 6). Additionally, economic development is represented by some governments as a method for mitigating or preventing political conflict.

After the fall of the Soviet Union in 1991, US Congress began to question the necessity of international aid and development, which was seen as a method for reducing the expansion of Soviet-style socialism during the Cold War. In an effort to remain relevant, organizations such as USAID reconceptualized economic development as a method for preventing conflicts globally (Duffield 2001). This reconceptualization viewed aid and development as a method for expanding markets and liberalizing economies to increase incomes and avoid scarcity, which was viewed as a symptom of civil unrest and conflict (Essex 2013). With the collapse of the Soviet Union in 1991, funding for socialist economic development programs decreased significantly, leaving spaces open for additional capitalist development schemes.

However, market-driven neoliberal capitalism has widened rather than narrowed the economic divisions between places and people. These distances became hyper-visible during and in the aftermath of disasters, both environmental (such as earthquakes, floods or tsunamis) and geopolitical (such as armed conflict). Aid and development have become geopolitical methods for wealthier countries to influence (or extract resources from) countries with less access to material resources, particularly when they are further weakened by a disaster. Disasters in wealthy countries, such as the United States, have been used as opportunities to redevelop areas and marginalize or remove impoverished communities from certain geographic areas. The redevelopment of New Orleans after Hurricane Katrina (particularly the ninth ward, where a majority of lower-income African American families lived) exemplifies this type of redevelopment, which focuses on economic opportunity rather than the return of displaced residents (Klein 2007). The areas hit hardest by the hurricane and the lack of assistance to poor and racially marginalized residents (from US agencies including Housing and Urban Development) illustrates the segregated histories of the United States, "where housing for black, working-class communities [is] located in the least desirable areas, with limited employment, social services and amenities" (Braun and McCarthy 2005).

Humanitarian relief and economic development assistance follow in the wake of disasters and destruction of places and property. Environmental disasters highlight different experiences of vulnerability. Some people are much more vulnerable than others during and in the aftermath of disaster. Gender as it intersects with race, class and other social categories remains crucial to our understanding of the interrelationship between environmental and social worlds. Therefore, examining aid and development through the lens of intersectional gender identities during and in the aftermath of political, social and environmental disasters reveals various experiences of inequality, dispossession, abuse and marginalization.

Disaster capitalism

Scholar and public intellectual Naomi Klein (2007) defines capitalist responses to geopolitical and environmental disasters as a platform for economic and political restructuring. Disaster capitalism suggests a reorganization of places and people as integral to reconstruction efforts particularly when disorganization, lack of governance and social chaos facilitate the introduction of neoliberal capitalism through humanitarian assistance and economic development. Aid and development organizations compete for access to locations beset by disaster in order to set up operations and access resources (Cooley and Ron 2002). Competition among NGOs for access to geographic space for offices and living quarters was particularly pronounced after the 2010 earthquake in Haiti.

The UN, as the largest multinational aid organization, did not adequately manage this situation and has been repeatedly cited for poor management which increases the vulnerabilities of various populations in need of humanitarian assistance. "John Holmes, the coordinator of the UN Office for the Coordination of Humanitarian Affairs, criticized his employees for failing to adequately manage the relief effort in Haiti after the quake" (Pierre-Louis 2011, 200). Additionally, the UN transferred personnel from Nepal to Haiti to assist as peacekeepers, and these UN employees brought the disease cholera with them, which spread quickly and killed thousands of individuals. In 2016 the UN accepted responsibility for its role in this outbreak; however, class action lawsuits against the UN were thrown out of court, because the UN invoked its legal immunity under the Convention on the Privileges and Immunities of the United Nations (Pilkington 2016, Pillinger, Hurd and Barnett 2016).

Gender-based exclusions for assistance occur at multiple levels in the wake of disasters, but disasters can also be opportunities to address gender inequalities (Horton 2012). For example, in 2004 Sri Lanka was embroiled in both a civil war between the government and the Liberation Tigers of Tamil Eelam (LTTE) and recovery efforts following the devastating tsunami in December of that year. The majority of victims of the tsunami were women because of existing gendered divisions of labor (Hyndman 2011). Because women's labor as merchants was located along the seashore in larger numbers than men, they were in the path of the tsunami's destruction and therefore disproportionately injured and killed compared to men whose labor or soldiering placed them further inland (Hyndman 2011).

While feminist scholars identify that men's and women's experiences of disasters are often different, it is necessary to address the intersection of social, racial, economic and political relations, which reveal inequalities in disaster management (Seager 2014, Hyndman 2011). "A feminist approach examines how gender differences produce material, social, and other inequalities between the sexes, but also how other social locations (such as caste, class, and ethno-national identity) produce disparities and marginalization" (Hyndman 2011, 63). If the needs of individuals and groups are to be adequately met during times of uncertainty, then scholars and policymakers must examine intersectional gender identities in spaces of environmental disaster and conflict. Hyndman's (2011) discussion of the "dual disasters" of political conflict and the tsunami in Sri Lanka shows how aid and development policies can entrench hierarchical relations among groups of people. She argues that "well-intentioned work of development staff and humanitarian actors can unwittingly reproduce and perpetuate existing gender, racial, and geographical hierarchies by uncritically promoting certain kinds of projects" (Hyndman 2011, 77). For example, these projects reinforce the existing power of elites, leaving already vulnerable populations further marginalized from assistance and without access to economic opportunity. Therefore, the political use of aid and development promotes certain modes of political and civic engagement that can and often do cause problems and disruptions that make it more (rather than less) difficult for vulnerable populations to access needed resources.

In an effort to avoid the politicization of assistance, aid organizations in Sri Lanka attempted to separate humanitarian and political spaces. This was premised on the idea that humanitarianism should rise above politics and attend to acute human needs for basic survival irrespective of political affiliation. However, in practice humanitarian spaces remain highly politicized (Kleinfeld 2007), thus, challenging neutrality as the guiding philosophy of humanitarianism. Therefore, once humanitarian organizations decide which individuals are worthy (or not) of assistance, the provisioning of assistance becomes divisively political. Humanitarian organizations often play a role (whether intentionally or not) in politicizing assistance by designating individual bodies as more or less worthy of assistance. The politicization of assistance can be corrosive and often exacerbates existing tensions and conflicts within a given society, which in some cases has led to violent attacks against aid and development organizations. Thus, humanitarian and development organizations operating in situations of conflict increasingly rely on private security and military corporations to ensure the safety of international workers (Fluri 2011a). The militarization of security (from governments and private corporations) has become integral to peace building and humanitarianism during conflict and in the wake of disaster. The following section examines political conflict, gender, aid and economic development followed by the role of international organizations in mitigating and perpetuating conflicts under the banner of assistance.

Gender and vulnerability

Internationally, the United Nations has identified the inherent vulnerabilities women and children experience during armed conflict. In an effort to address and mitigate gender differences during war, the United Nations Security Council initiated resolution 1325 (UNSCR 1325) on women, peace and security, which was adopted in 2000.

> The resolution reaffirms the important role of women in the prevention and resolution of conflicts, peace negotiations, peace building, peacekeeping, humanitarian response and in post-conflict reconstruction and stresses the importance of their equal participation and full involvement in all efforts for the maintenance and promotion of peace and security. Resolution 1325 urges all actors to increase the participation of women and incorporate gender perspectives in all United Nations peace and security efforts. It also calls on all parties to conflict to take special measures to protect women and girls from gender-based violence, particularly rape and other forms of sexual abuse, in situations of armed conflict. The resolution provides a number of important operational mandates, with implications for Member States and the entities of the United Nations system.
>
> (UN 2000)

UNSCR 1325 has in some respects been revolutionary because it calls international attention to gender-based violence during (and in the aftermath of) conflict. It followed the International Criminal Tribunal for the former Yugoslavia (1993) and the International Criminal Tribunal for Rwanda (1994), which identified the use of rape as a weapon during political conflict as a war crime and crime against humanity. Critiques of UNSCR 1325 point out the generalizations made about women, which reinforce binaries about gender and violence, i.e., positioning women as victims and objects of violence (Shepherd 2008). UNSCR 1325 and other international laws are limited by the inability of international organizations to monitor, evaluate and enforce these resolutions without significant support from powerful countries. Additionally, the UN (and other assistance organizations) cannot always ensure security, particularly when the UN's own personnel are involved in abuse and exploitation. Furthermore, civilian noncombatant men rarely, if ever, occupy the status of victim, even when they are targets of sexual violence, war crimes and other corporeal violations, despite efforts to include them in subsequent UN Security Council resolutions (Carpenter 2006). Sexual violence against women and men during conflict has been used as an emasculating weapon against the "enemy" and a form of intimate torture for the victims. The political use of sexual abuse and rape illustrates how geopolitical violence is intimately experienced at an individual scale. Thus, vulnerabilities become a space upon which geopolitical actions are articulated. Government-sponsored militaries remain legitimate arbiters over the means of violence, including war-related conflicts. The violence associated with war reverberates well beyond the killing of soldiers and civilians. War disrupts infrastructures and leaves enduring injury upon people and places. For example, more people die from disease and lack of resources, including healthcare, than they do from violence in most contemporary wars.

Gender and age are two methods to galvanize and generate financial or institutional support for aid in the aftermath of disaster. While women's and children's vulnerabilities are often more acute and different than those of able-bodied men, these vulnerabilities can be manipulated to reinforce certain political and economic agendas. "Analysis . . . indicates that although women are more often more vulnerable to disasters than men," we must also remember that "they are not just the 'helpless victims', as often represented" (Ariyabandu 2009, 7). Women have largely been marginalized or underrepresented in formal organizations tasked with emergency planning and relief, particularly in decision-making capacities. However, "women outnumber men" at the scale of communities and informal disaster relief measures (Ariyabandu 2009, 7). While formal and informal methods of assistance and care may be gendered, both are valuable and essential methods for provisioning aid. Informal methods of care within communities are often more effective and long-term than formal systems, which can be stymied or slowed by bureaucracy, politics and official hierarchies. Conversely, formal aid provisions can provide funds and resources not readily available to local and informal forms of assistance in the wake of disaster.

Vulnerable populations are also at risk of physical and sexual violence, human trafficking and ill health in situations of conflict and disaster. Lawlessness and chaos occur during and in the aftermath of environmental catastrophe or political conflict. In other cases, civilian populations on the receiving end of assistance – particularly in situations of conflict or disaster mitigation – continue to be the targets of abuse by the very aid workers and peacekeepers entrusted with attending to their needs (Aoi, de Coning and Thakur 2007, Vandenberg 2005). The UN's edited volume *Unintended consequences of peacekeeping operations* identifies several examples of abuse and exploitation suffered by vulnerable and displaced populations who were fleeing conflict (Aoi, de Coning and Thakur 2007). Examples include human and sex trafficking or requiring sex acts in exchange for food or other resources. In response, the UN increased the number of female peacekeepers in an effort to change the gender imbalances of peacekeepers and to prevent abuse by male peacekeepers and aid/development workers. Altering the gender of peacekeepers relies on assumptions about men and women regarding violence and abuse that may be inaccurate, and does not address the problem – caused by power imbalances – namely the power aid workers have over vulnerable populations in need of assistance.

Some organizations expect an increase in female workers (from peacekeepers to aid and development workers) to decrease or mitigate abuse, gender-based violence and trafficking orchestrated by international workers. While these attempts are touted as methods for increasing the number of women within these professions, using women to improve male behavior rests on several gendered assumptions. For example, gender-based solutions assume that women are uniquely suited to protect other women and that men are not sufficiently able or willing to provide the same protections. This assumption is problematic because it uses gender as the main category for deciding who can and cannot offer appropriate protection and it does not identify men as allies in the struggle against gender-based violence (Jennings 2011).

Large aid/development missions – such as in the immediate aftermath of the US-led invasions of Afghanistan (2001) and Iraq (2003) – have included influxes of thousands of international aid and development professionals, who have a significant but not always positive impact on local places and people (Fluri and Lehr 2017). For example, sex workers, brothels and human trafficking commonly increase after the onset of international assistance. In Liberia and the Democratic Republic of Congo, "survival prostitution and trafficking" occurred in peacekeeping sites (Jennings 2011, 6). Additionally, despite the UN's moratorium on workers having sexual relationships with local beneficiaries and their zero-tolerance policy on sexual abuse, enforcement remains largely insufficient (Jennings 2011). In 2017 there were two different reports of child sexual abuse by UN peacekeepers/aid workers in the Democratic Republic of Congo and more than 100 UN peacekeepers ran a child sex ring in Haiti for over ten years, without prosecution. Human trafficking, sexual abuse and an increase in sex work nearly always accompany the onset of a large post-disaster aid/development mission (Jennings 2011).

With the sudden flood of international aid/development professionals from a variety of governments and NGOs throughout the globe to mitigate the effects of disasters, an auxiliary economy is formed to meet the needs, wants and desires of international workers. This includes a number of service employees, from domestic workers to security personnel, and sites of leisure, from bars to brothels. International workers sometimes argue that sex work is a necessary part of working in a conflict zone, believing the false narrative that it keeps men from raping female co-workers because sex workers provide them with the opportunity to release sexual tension (Fluri 2011b). These assumptions about sex workers and other female aid/development workers highlight a particular type of male privilege that positions the male desire for sex as a "need" that outweighs the rights and needs of others, mostly female persons. This also situates a racist, classist and dichotomous separation between "legitimate" female international aid/development workers (mostly white and socioeconomically privileged) and "delegitimate" female international sex workers (mostly nonwhite and socioeconomically disadvantaged).

Female aid/development workers also engage in sexual relationships with coworkers and citizens of their host countries. For example, in Afghanistan, some international female workers engaged in sexual intimacy with Afghan men. In most cases these men were employees of their female sex partners (i.e., drivers, security guards or office workers). While these sexual relationships may have been consensual, they were not equal; female international workers had more economic power and mobility, and Afghan men retained a dominant position in the local patriarchal hierarchy (Fluri 2011b). Jennings (2014) calls for doing peacekeeping differently by addressing the difficulties and needs of both peacekeepers and the populations they serve by creating and enforcing fair labor practices for peacekeepers and subcontractors. "Reconsideration of the often chauvinistic, gendered messages about locals that are institutionally endorsed and disseminated through training and security regimes remains an important step in addressing abusive practices" (Jennings 2014, 14). Addressing sexual desire and other forms of intimacy within spaces of insecurity and uncertainty should be openly discussed and addressed by international organizations in order to change organizational cultures that situate vulnerable populations as ripe for further exploitation.

Humanitarian militaries and liberal peace building

Contemporary political conflicts are fought in populated spaces, including urban centers. Thus, battle zones and civilian spaces have become largely indistinguishable (Hyndman and De Alwis 2004). Therefore, reconstruction, humanitarian assistance and economic development can and do occur during and in the aftermath of war. Victory may be declared politically while armed conflict continues between different factions. War and related political conflicts devastate infrastructures and damage places necessary for basic human survival. Physical violence during war is integral to other forms of human security (Roberts 2008). People living within conflict spaces experience illness and death from violence and related infrastructure collapse, e.g., lack of potable water, shelter, food and adequate clothing (Roberts 2008). Men's and women's experiences of conflict and related injuries are significantly different and vary across the life cycle, i.e., children, the elderly and disabled individuals remain more vulnerable to illness and lack of proper food, water and shelter than able-bodied men and women.

The US-led coalitions in Iraq and Afghanistan changed course after initial "smart" bombing campaigns to standing armies, counterinsurgency and drone technologies. Winning the hearts and minds of local populations became integral to US-led counterinsurgency (COIN) techniques. Therefore, military personnel provided humanitarian services from medical care to road and bridge building in an effort to garner the favor of local populations. This has led to an increased blurring of the lines between military action and humanitarian or development assistance. Gender has been incorporated into the ways in which the US military orchestrates the use of humanitarian assistance as part of "winning hearts and minds" campaigns, particularly in Iraq and Afghanistan. The development of Female Engagement Teams (FETs) is an example of the ways in which gender has become an integral geopolitical tool that manipulates humanitarian aid or development assistance as a counterinsurgency strategy. The political uses of humanitarian aid more so than development directly challenge the ideal of neutrality as a guiding (but rarely followed) tenet of humanitarian assistance.

FETs were conceptualized by the US military in an attempt to address concerns among civilians in Afghan and Iraqi communities about house raids and the need to include women's perspectives in military intelligence gathering. FETs are also mistakenly associated with contributing to Afghan women's empowerment (Dyvik 2014). The FET philosophy seeks to use female military personnel to entice Afghan men to enter public space with the expectation that they will be curious about these soldiers. Additionally, FETs offer "humanitarian" services such as medical care in an effort to draw villagers out of their homes, in order for the military to gather information about insurgents (Dyvik 2014). Representations of the FET program by the US military and Department of State situate female soldiers as engaging in acts of humanitarian assistance and care work rather than in acts of violence, torture or brutality (Fluri 2015). Thus gender, particularly the femininity of these soldiers – and their assumed association with care work – is manipulated by the US military in an effort to win the hearts and minds of local civilians through its COIN operations (Dyvik 2014, Fluri 2015).

Within the policy and enactment of counterinsurgency, "a new form of masculinity emerges, authorized by consumerism and neo-liberal feminism, in which 'manliness' is softened, and the sensitive masculinity of the humanitarian soldier-scholar (white, literate, articulate, and doctorate-festooned) overshadows the hyper-masculinity of warrior kings" (Khalili 2011, 5). Humanitarian military influences "includes a significant proportion of female officials, is coded as feminine work, and enjoys strong and visible advocacy," thus geopolitically mapping and manipulating gender (Khalili 2011, 5). Women have been particularly essential in the creation and maintenance of the so-called "humanitarian" components of the US-led global War on Terror. For example, Montgomery McFate, an anthropologist, conceptualized the Human Terrain System (HTS), which employs anthropologists and ethnographic techniques in an effort to infiltrate and control civilian populations (Khalili 2011). Attempting to win over civilian populations through the militarization of social science methods proved to be an ineffective counterinsurgency or anthropologic method, and HTS program operations were closed in 2014. Other socioeconomic development methods have been used in an effort to mitigate political conflict and build peace. Mitigating conflict by crafting liberal peace is another method used to transition from conflict to economic and political stability.

Liberal peace

Liberal peace merges and mingles the term "liberal" (in both economic and political regards) with the term "peace" (focused on resolving conflicts and reconstructing societies) (Duffield 2014). Liberal ideologies on morality and the universality of human rights abound with the paradoxes of militarized aid and development (Slaughter 2007). In many conflict zones an intersecting web of connections exists between institutions tasked with promoting liberal peace through aid/development and those that profit from these operations (see Chapter 3). The integration between aid/development and private or government security forces/militaries, along with the growing use of humanitarian assistance by militaries, leads to a much more tangled connection. On-the-ground aid and development organizations operate in active conflict zones (Goodhand 2006). Liberal peace building aims to change broken governments and societies affected by war into secure and cooperative countries (Duffield 2014). Rather than work with or for the local government, these organizations attempt to sustain civil society, thus promoting and often expecting local and international NGOs to maintain and sustain civil society. Aid/development policy attempts to prevent and mitigate conflict by crafting the idea of liberal peace through economic development (Mac Ginty and Williams 2009). Meanwhile, more effective aid policies are considered necessary for reducing conflict and therefore reducing chronic poverty (Goodhand 2003). Liberal peace ideologies fail partly due to the increasing reliance on private security, military, logistics and construction companies to foster these programs. Aid and development organizations are often involved in the perpetuation of conflict through their relationships with military and security forces, as well as attempting to reduce and ultimately end the conflict (Goodhand 2006). Therefore, efforts to develop a liberal peace through economic development remain stymied by the growing interconnectedness between militaries and aid/development organizations. Additionally, the continued conflicts in Afghanistan, Iraq and Syria have contributed to an extensive humanitarian crisis associated with individuals fleeing conflict and seeking refuge and resettlement.

Displacement

Disasters and conflict often result in the displacement of many individuals from their homes and communities. **Internally displaced persons** (IDPs) are individuals who remain within their own country but have been displaced from their homes. In 2017, there were a total of 65.3 million forcibly displaced persons worldwide; 21.3 million were refugees, 10 million were stateless people and only 107,100 refugees were resettled. According to the International Displacement Monitoring Centre, this is equivalent to one person fleeing by force every second (IMDC 2017). Refugees live in similarly bounded and vulnerable circumstances. Hyndman and Giles (2016) argue that refugee camps should not be referred to as communities but rather as "a non-community of the excluded" who are managed by international governmental and nongovernmental organizations.

Refugees experience extensive and difficult forms of mobility when they move from their homes to refugee camps. Many refugees experience multiple forms of displacement until they are finally settled in a new country of residence (or are able to return home). Other refugees remain in refugee camps for extended periods of time, and according to the United Nations High Commissioner on Refugees, the average stay in a refugee camp is 17 years. (Some individuals are born and live most or all of their childhood in a refugee camp or as an IDP.) Displacement as an outgrowth of disaster and conflict has become a humanitarian and development issue globally. Much of the immediate assistance provided to individuals fleeing conflict or disaster takes the form of humanitarian aid, while the extensive needs of refugees and migrants become part of longer-term development projects. Due to the protracted conflicts in Afghanistan, South Sudan, Somalia, Myanmar and Syria, approximately 68 percent of refugees come from these five countries. The countries that hosted the most refugees in 2017 were Turkey (3.5 million), Pakistan (1.4 million), Uganda (1.4 million), Lebanon (998,900), Iran (979,400), Germany (970,400) and Bangladesh (932,200) (UNHCR 2018).

In comparison, the United States resettled 33,400 refugees in 2017 (UNHCR 2018), and in the same year the European Commission promised to resettle 50,000 refugees over two years. These numbers reveal the role of spatial proximity for countries absorbing refugees who are fleeing bordering countries, along with the global inequalities associated with resettlement, where the wealthiest countries' absorption of refugees remains fractional compared to countries with far less wealth and fewer resources. Additionally, gender remains a significant factor in determining a person's worthiness for resettlement. Young men fleeing conflict or forced military conscription (such as in Syria) are often viewed as a potential threat rather than a vulnerable group based on gendered and racist assumptions about able-bodied Muslim men. Women and children are more likely to be viewed as vulnerable. However, all refugees need to prove their vulnerability in order to receive asylum or establish permanent residency in another country. The following two case studies, Afghanistan and Syria, are presented to provide more context about these conflicts and the humanitarian and development crises encountered by the citizens of these countries.

Case studies: Afghanistan and Syria

The first case study examines the way in which gender was used as a legitimate reason for continued war, occupation and aid/development in Afghanistan. The use of gender as a tool of geopolitics has been evident through various phases of this US-led war. The second case study, which focuses on the Syrian civil war, explicates the ways in which populations, such as young men attempting to avoid military conscription, are viewed as potential threats rather than vulnerable people fleeing conflict.

Afghanistan

In response to the attacks against the United States on September 11, 2001, the US led an international military invasion of Afghanistan on October 7, 2001. This military action sought to find Osama bin Laden (the leader of Al Qaeda and architect of the attacks against the United States) and remove the Taliban from power in Afghanistan. The Taliban were notorious for their strict control over Afghan society and their removal of women from public space (they banned women from working, prevented girls from attending school, and placed restrictions on women's mobility). Prior to 9/11, many transnational feminist organizations along with Afghan feminist organizations had sought assistance to resist the Taliban government. In the aftermath of the US-led invasion of Afghanistan, geopolitical discourse shifted from the hunt for bin Laden to saving Afghan women. Many feminist scholars and activists remained wary of this geopolitical concern for Afghan women, and highlighted the ways in which women's rights were co-opted for military and political purposes (Hunt 2002, Abu-Lughod 2013). Despite these critiques, development efforts from the United States and other countries and NGOs created programs and projects to save, assist and aid Afghan women and girls.

The international aid/development focus on women in some respects provided new opportunities for Afghan women's education, employment, mobility and participation in politics. While geopolitical and aid/development discourse focused on supporting, assisting, liberating and empowering Afghan women, funds for projects dedicated to women received much less funding than other aid/development allocations. In some cases, women-specific funding led to backlashes from men who perceived these programs as contributing to male unemployment and concerns about foreign influences on home and family life (Abirafeh 2009). These projects did not attempt to integrate reforms within existing gender regimes or belief systems. In many cases local feminist and women's rights activists' ideas for improving women's lives were not incorporated into internationally funded projects based on Western-liberal and capitalist ideologies, which repeatedly ran counter to local sociocultural mores (Kandiyoti 2007).

For example, Afghan widows were lauded by internationals both during and after the Taliban regime for their ability to move in public space without male accompaniment (protection) and to work outside the home during the Taliban regime. Many of the ways in which Afghan widows were represented by internationals identified their work as resistance to patriarchy, romanticizing them as rebelling against the Taliban (Abu-Lughod 1990, Daulatzai 2006). These assumptions discounted the real-life vulnerabilities and difficulties experienced by these women. For example, a widow's ability to move in public space without male accompaniment does not take into consideration the lack of protections and insecurity this mobility signifies or the actual experience of widows. International organizations often positioned Afghan women in opposition to Afghan men or the patriarchal social order, without considering that international intervention can be extensively problematic and create even more difficulties for women (Daulatzai 2006).

In the early intervention period (2002–08), many organizations (both for- and not-for-profit) attempted to assist Afghan women through income-generating projects. Some projects attempted to capitalize on the geopolitical expectation that the United States and its allies were "saving" Afghan women. Therefore, they used the US-led geopolitical discourse of saving/assisting Afghan women from the Taliban to market items "made by Afghan women."

> In these marketing scenarios one does not merely purchase a handbag, scarf, or shirt made in Afghanistan; one buys the tale of a woman in the process of being liberated. Through the act of consumption, customers enjoy a "feel good" moment associated with their virtual participation in the alleviation of an Afghan woman's suffering toward economic liberation.
>
> (Fluri and Lehr 2017, 57–58)

Women's labor was therefore translated as liberation through effective advertising and marketing, while their incomes remained low and their work remained unregulated and irregular.

Similar to the discussions in Chapter 3, consumption is viewed as a form of assistance. This presents a dangerous precedent because it reduces the alleviation of suffering to the temporal limitations of fashion trends. As Thrift (2008, 14) argues, "style wants us to love it and we want to be charmed by it." Thus, marketing "Afghan woman made" products through the lens of assisting, saving, empowering or liberating women reduces their often difficult and low-wage labor to the temporally limited trendiness of stylized assistance. It further charms the consumer to purchase these items as an act of virtual benevolence. Purchasing an Afghan woman made product is intended to captivate "the consumer by suggesting this purchase and product are an expression of 'doing good'" (Fluri and Lehr 2017, 63).

The impetus to save and assist Afghan women did not take into consideration the needs, desires and diversity of Afghan women themselves. Similar to Afghan men, the women are an extensively diverse group of disparate ethnicities, belief systems, socioeconomic class levels, educational experiences and skills. For example, urban educated women's experiences are often dramatically different from those of rural women who don't have the same access to education. Thus, the geopolitical use of gender focused an extensive amount of attention, funding and programing on women in Afghanistan, while a review of these efforts from the Special Investigator General for Afghanistan Reconstruction (SIGAR) found *no* conclusive evidence that USAID and the US Department of State programs designed to assist Afghan women had made an appreciable improvement in their lives (Sopko 2014).

The protection of women, children and homes remains integral for civilian populations whose lives are caught in the crossfire of contemporary political conflicts. In many cases military strategy runs counter to protecting the sanctity and security of the home. For example, US-led coalition forces in Afghanistan and Iraq implemented night raids on homes in an effort to infiltrate insurgent forces and prevent them from seeking shelter in civilian domestic spaces. However, these house searches were perceived as an affront to the sanctity of the household and by definition a violation of women, because the home is expected to be a private and secure space. The magnitude of breaching the threshold of the home without permission has been identified as akin to violating a sovereign border and as devastating to families as civilian causalities (Fluri 2011a).

The discourse of saving and assisting women did not match the realities of women's experiences. Afghan women and men need assistance; however, assistance should be provided within an acceptable sociocultural context. Additionally, in order for assistance to be effective and development sustaining, projects and programs must be designed to incorporate the diverse and disparate needs of women (and men) across intersecting social, ethnic, economic and political categories.

Syria

The Syrian conflict and subsequent refugee crisis continue to severely limit many Syrians' ability to live without conflict or fear of harm. The United Nations High Commissioner for Refugees (UNHCR) has developed a regional response plan in an effort to coordinate efforts among different UN agencies and NGOs. However, this plan has had little effect because of the enormity of displaced persons and continued political and economic tensions regionally (Freedman, Kivilcim and Baklacioğlu 2017). This refugee crisis can be classified as one of "chronic vulnerability," particularly for individuals who are not registered with the UN or have lost their registration status and live in a type of limbo in neighboring countries (Zetter and Ruaudel 2014, 7).

> Socio-economic factors and the lack of legal status increase refugees' susceptibility to a range of human rights abuses and vulnerabilities whether in camps or urban settings. Forced and early marriages have reportedly risen compared to the pre-crisis period and incidents of domestic violence, sexual and gender-based violence and violence against children are high.
>
> (Zetter and Ruaudel 2014, 9)

While these gendered issues are imperative to examine, we must do so within the context of war and survival. For example, early or forced marriages may in some examples be against the autonomous rights of women; however, in other cases they may be used by parents to ensure their daughter(s) will be protected and provided for, particularly in the event of a parent's death (Charles and Denman 2013). Additionally, legal measures in host countries that prevent or prohibit arranged, polygamous or forced marriages actually generate deeply negative consequences for the women they claim to protect (Eichenberger 2012, Freedman, Kivilcim and Baklacioğlu 2017). Therefore, the abusive or negative treatment women experience has been largely attributed to Syrian "culture," when they are often the direct result of trauma caused by conflict and economic and political uncertainty.

Many single men have secretly fled Syria since March of 2012, when the Syrian government banned men (between the ages of 18 and 42) from leaving the country without formal authorization, in order to force them into military conscription (Davis, Taylor and Murphy 2014). Single men have simultaneously become viewed as less vulnerable and potentially threatening, which has made it more difficult for them to seek and receive assistance in host countries. The assumption that single men are by definition a potential threat, and therefore invulnerable,

> may serve to increase the vulnerability of these single men, who are forced to become illegal residents in the host country and who, in addition, may find themselves last on the list for humanitarian aid, behind those deemed more "vulnerable" by the international organizations and aid agencies.
>
> (Freedman, Kivilcim and Baklacioğlu 2017, 6)

Additionally, within Syria during the height of the conflict, many international aid organizations were not able to safely function in certain areas, leading local Syrian women to actively distribute provisions of medicine and food (Haddad 2014). These examples illustrate the importance of accounting for men's vulnerability and women's active role in their communities, providing for their families and protecting others. Due to conflict-based displacements, many women are now heads of their households, and are tasked with both domestic responsibilities and generating income.

While there are many forms of gendered vulnerability, care, assistance and mobility associated with the Syrian refugee crisis, a one-size-fits-all gender analysis will not address the acute and enduring needs of women, men and children. Both the promise and problems of gendered attention to a crisis of this magnitude have the tendency to map the entire population of refugees with a rigid rather than fluid understanding of gendered experiences. The bureaucratic makeup of aid organizations does not often allow for nuance when managing large numbers of displaced persons, which unfortunately reinforces rather than counters narrow conceptualizations of gender, need, vulnerability and capacity. Additionally, lesbian, gay, bisexual, trans and intersex (LGBTI) individuals experience barriers to basic services because of existing prejudices within their home and host communities (Freedman et al. 2017). A person may be rendered vulnerable due to conflict or disaster, both because of and in spite of one's gender. While women and children often experience acute vulnerabilities associated with age and gender, many men also experience extensive vulnerability, particularly civilian noncombatant men. For example, in the midst of armed conflict, a female soldier may experience far less defenselessness than an unarmed male civilian. Policy and bureaucratic procedures for addressing the enormity of corporeal vulnerabilities and displacements rely on aggregated analyses of populations, which leave little room for addressing the intersections of gender, sexuality, race and class to understand the complexities of vulnerability, need and assistance within and in the aftermath of conflict.

Future challenges

The critical examinations of aid/development discussed in this chapter explicate the ways in which assistance can "inflame conflict, degrade environmental conditions and have profoundly negative social and cultural consequences" (Mac Ginty and Williams 2009, 5). While some good comes from aid and development, it remains an uneven process that does not always benefit vulnerable groups or societies. Individuals who experience intersectional gender marginalization, or violence to their bodies because of the way they look, live or love, experience additional vulnerabilities in the aftermath of disaster. Additionally, able-bodied men can and do experience vulnerabilities that are not recognized by regulatory agencies because of their gender, physical abilities and expected status. There are several problems associated with international assistance missions, one of the largest being the abuse of vulnerable populations by peacekeepers or aid workers tasked with assisting these populations. Mitigating and stopping the abusive behavior of international aid workers remains an ongoing challenge that international and supranational organizations (such as the UN) must take seriously in order to stop this exploitation.

Additionally, waste of funds by international organizations severely limits the effectiveness of assistance in many places. Similarly, the role of politics in the process of providing relief in the wake of disasters and conflicts through humanitarian or development assistance does little to ensure adequate and fair distribution of needed resources. Political discourse can also sway public opinion or understanding about entire populations. For example, the global War on Terror has discursively represented Muslim women intersectionally, as oppressed by both Islam and local forms of patriarchy. However, the oppressions women experience in majority Muslim countries have less to do with Islam and more to do with entrenched social, economic and political inequalities within these countries and globally (Abu-Lughod 2013). Assumptions about the causes of one's intersectionally driven vulnerabilities can be used as a geopolitical distraction in an effort to invade, occupy or influence politics within a region and globally. Disaster and conflict have also become opportunities for organizations and governments to initiate new forms of development and enroll desperate individuals as low-waged laborers into the global capitalist economy.

Finally, the organization of state borders throughout the globe has become an increasingly fervent issue of disagreement and even conflict among various states. Mitigating the flight from conflict and the management of refugees is one of the most vexed and severe issues associated with border management and control. Today, the world is experiencing the largest refugee and displaced populations since World War II. How various countries assist or deny the plight of refugees remains an issue of great debate. The control of/over borders is premised on the protection of citizens within a country. Therefore, ensuring citizens' security has been habitually used as a reason to limit the acceptance of refugees (from both conflicts and environmental disasters). Individual bodies and their vulnerabilities are classified (both explicitly and implicitly) based on a number of other factors (such as gender, race, class, sex/sexuality, religion and political affiliation). For most individuals, the ability to cross borders to safety in a second or third country requires a tremendous amount of financial and social capital. Many borders have become more and more policed and militarized since the onset of the global War on Terror in 2001, while the number of people seeking assistance, shelter and refuge continues to grow. Several international organizations have organized around challenging the existence and management of borders in an effort to call attention to human-to-human rather than state-mitigated assistance (e.g., Doctors without Borders and Engineers without Borders). While these organizations are not without their faults, they have attempted to address acute human needs beyond or in lieu of government intervention.

Recommended reading

Games without rules: the often-interrupted history of Afghanistan, Tamin Ansary; We crossed the bridge and it trembled: voices from Syria, Wendy Pearlman

Recommended viewing

Whistleblower; Pray the devil back to hell; City of ghosts; Motherland Afghanistan

Questions for discussion

Locate active conflict areas throughout the globe. How many people have been negatively affected by conflict globally? Most people living in situations of conflict or in the aftermath of disasters become sick or die because of a lack of resources (food, water, shelter) than from violence. What daily resources are necessary to sustain your life? If you did not have access to these resources, what would you do? How would you ensure that you and your loved ones were able to survive? The organization Médecins Sans Frontières (Doctors without Borders) began an interactive simulation of refugee experiences, forcedfromhome.com. Examine the forcedfromhome.com site and reconstruct a similar exhibit at your university or within your community.

References

Abirafeh, L. (2009). *Gender and international aid in Afghanistan: the politics and effects of intervention*. London: McFarland and Company, Inc. Publishers.

Abu-Lughod, L. (1990). The romance of resistance: tracing transformations of power through Bedouin women. *American Ethnologist*, 17(1): 41–55.

Abu-Lughod, L. (2013). *Do Muslim women need saving?* Cambridge, MA: Harvard University Press.

Aoi, C., de Coning, C., & Thakur, R. (Eds) (2007). *Unintended consequences of peacekeeping operations*. New York, NY: United Nations University Press.

Ariyabandu, Madhavi M. (2009). "Sex, gender and gender relations in disasters." in E. enarson & P.G. Dhar Chakrabarti (Eds) *Women, gender and disaster: global issues and initiatives*, London: Sage Publications (pp. 5–17).

Braun, B.P., & McCarthy, J. (2005). Hurricane Katrina and abandoned being. *Environment and Planning D: Society and Space*, 23(6), 802–9.

Carpenter, C.R. (2006). Recognizing gender-based violence against civilian men and boys in conflict situations. *Security Dialogue*, 37(1), 83–103.

Charles, L., and Denman, K. (2013). Syrian and Palestinian refugees in Lebanon: the plight of women and children. *Journal of International Women's Studies*, 14(5), 95–111.

Cooley, A., & Ron, J. (2002). The NGO scramble: organizational insecurity and the political economy of transnational action. *International Security*, 27(1), 5–39.

Daulatzai, A. (2006). Acknowledging Afghanistan notes and queries on an occupation. *Cultural Dynamics*, 18(3), 293–311.

Davis, R., Taylor, A., & Murphy, E. (2014). Gender, conscription and protection and the war in Syria. *Forced Migration Review*, 47, 35–49.

Duffield, M. (2001). *Global governance and the new wars: the merging of development and security*. London: Zed Books.

Duffield, M. (2014). *Global governance and new wars: the merging of development security*, 2nd edition. New York and London: Zed Books.

Dyvik, S.L. (2014). Women as "Practitioners" and "Targets": gender and counterinsurgency in Afghanistan. *International Feminist Journal of Politics*, 16(3): 410–29. https://doi.org/10.1080/14616742.2013.779139.

Eichenberger, S.L. (2012). When for better is for worse: immigration law's gendered impact on foreign polygamous marriage. *Duke Law Journal*, 61, 1067.

Essex, J. (2013). *Development, security, and aid: geopolitics and geoeconomics at the U.S. Agency for International Development*. Athens, GA: University of Georgia Press.

Fluri, J.L. (2011a). Bodies, bombs and barricades: geographies of conflict and civilian (in)security. *Transactions of the Institute of British Geographers*, 36(2), 280–96.

Fluri, J. (2011b). Armored peacocks and proxy bodies: gender geopolitics and aid/development spaces of Afghanistan. *Gender, Place & Culture*, 18(4), 519–36.

Fluri, J.L. (2015). "Feminist political geography." In J. Agnew, V. Mamadouh, A. Secor, & J. Sharp (Eds) *The Wiley Blackwell companion to political geography*. Oxford: Wiley Blackwell (pp. 235–47).

Fluri, J.L., & Lehr, R. (2017). *The carpetbaggers of Kabul and other American-Afghan entanglements: intimate development and the currency of gender and grief*. Athens, GA: University of Georgia Press.

Freedman, J., Kivilcim, Z., & Baklacioğlu, N.O (2017). "Introduction: gender, migration and exile." In *A gendered approach to the Syrian refugee crisis*, New York, NY: Routledge (pp. 1–15).

Goodhand, J. (2003). Enduring disorder and persistent poverty: a review of the linkages between war and chronic poverty. *World Development*, 3(3), 629–46.

Goodhand, J. (2006). *Aiding peace? The role of NGOs in armed conflict*. Boulder, CO: Lynne Rienner Publishers.

Grewal, I. (2014). "American humanitarian citizenship: the 'soft' power of Empire." In S. Ponzanesi (Ed.) *Gender, globalization, and violence: postcolonial conflict studies*, New York, NY: Routledge (pp. 64–81).

Haddad, Z. (2014). How the crisis is altering women's roles in Syria. *Forced Migration Review*, 47, 46–50.

Horton, L. (2012). After the earthquake: gender inequality and transformation in post-disaster Haiti. *Gender & Development*, 20(2), 295–308.

Hunt, K. (2002). The strategic co-optation of women's rights. *International Feminist Journal of Politics*, 4(1), 116–21.

Hyndman, J. (2011). *Dual disasters: humanitarian aid after the 2004 tsunami*. Sterling, VA: Kumarian Press.

Hyndman J., & De Alwis, M. (2004). Bodies, shrines, and roads: violence, (im)mobility and displacement in Sri Lanka. *Gender, Place & Culture*, 11, 535–57.

Hyndman, J. and Giles, W. (2016). *Refugees in extended exile: living on the edge*. London: Routledge.

IMDC (2017). *Global report on internal displacement*. Norwegian Refugee Council. www.internal-displacement.org/global-report/grid2017/. Accessed 2/16/2019.

Jennings, K.M. (2011). *Women's participation in UN peacekeeping operations: agents of change or stranded symbol*. Oslo: Norwegian Peacebuilding Resource Centre.

Jennings, K.M. (2014). Service, sex and security: gendered peacekeeping economies in Liberia and the Democratic Republic of the Congo. *Security Dialogue*, 45(4), 313–30.

Kandiyoti, D. (2007). Old dilemmas or new challenges? The politics of gender and reconstruction in Afghanistan. *Development and Change*, 38(2), 169–99.

Khalili, L. (2011). Gendered practices of counterinsurgency. *Review of International Studies*, 37(4), 1471–91.

Klein, N. (2007). Disaster capitalism. *Harper's Magazine*, 315, 47–58.

Kleinfeld, M. (2007). Misreading the post-tsunami political landscape in Sri Lanka: the myth of humanitarian space. *Space and Polity*, 11(2), 169–84.

Mac Ginty, R., & Williams, A. (2009). *Conflict and development*. London: Routledge.

Pilkington, E. (2016). UN makes first public admission of blame for Haiti cholera outbreak. *Guardian*, August 18. www.theguardian.com/world/2016/aug/18/un-public-admission-haiti-cholera-outbreak. Accessed 2/18/2019.

Pillinger, M., Hurd, I., & Barnett, M. (2016). How to get away with cholera: the UN, Haiti and international law. *Perspectives on Politics*, 14(1), 70–86.

Pierre-Louis, F. (2011). Earthquakes, nongovernmental organizations, and governance in Haiti. *Journal of Black Studies*, 42(2), 186–202.

Roberts, D. (2008). *Human insecurity: global structures of violence*. New York, NY: Zed Books.

Seager, J. (2014). "Disasters are gendered: what's new?" In Z. Zommers, & A. Singh (Eds) *Reducing disaster: early warning systems for climate change*, New York, NY: Springer (p. 265).

Shepherd, L.J. (2008). *Gender, violence and security: discourse as practice*. New York, NY: Zed Books.

Slaughter, J.R. (2007). *Human Rights, Inc.: the world novel, narrative form, and international law*. New York, NY: Fordham University Press.

Sopko, J.F. (2014). Afghan women: comprehensive assessments needed to determine and measure DOD, state, and USAID progress. Special Inspector General for Afghanistan Reconstruction (SIGAR), 15–24 Audit Report (December): 1–44.

Thrift, N. (2008). The material practices of glamour. *Journal of Cultural Economy*, 1, 9–23.

UN (2000). Resolution 1325, adopted by the Security Council at its 4213th meeting, on October 31, 2000. https://daccess-ods.un.org/TMP/9623053.07388306.html. Accessed 2/15/2019.

UNHCR (2018). Global trends: forced displacement in 2017. www.unhcr.org/statistics/unhcrstats/5b27be547/unhcr-global-trends-2017.html. Accessed 2/18/2019.

Vandenberg, M. (2005). "Peacekeeping, alphabet soup, and violence against women in the Balkans." In D. Mazurana, A. Raven-Roberts, & J. Parpart (Eds) *Gender, conflict and peacekeeping*, Lanham, MD: Rowman & Littlefield.

Zetter, R., & Ruaudel, H. (2014). Development and protection challenges of the Syrian refugee crisis. *Forced Migration Review*, 47, 6.

10 Alternative development and decolonization

Introduction

According to the environmental watchdog group Global Witness, 200 activists attempting to protect their land, water and natural resources were murdered in 2016, including Dolores Cáceres in Honduras. One murder is one too many when it comes to peaceful protest against land dispossession. The rate of murders of individuals protecting their land has escalated dramatically since 2010, as has imprisonment of indigenous activists such as those at Standing Rock, North Dakota, USA (see Chapter 4). The methods of securing resource extraction (murder, imprisonment) and appropriation of surplus (theft by taking, slavery) that modern economies claim to have condemned continue apace. Enrollment in capitalism through enclosure and dispossession is often a force few are strong enough to successfully resist, and too many pay with their lives and freedom. It is no accident that those who are killed and imprisoned are too frequently marginalized based on their gender, race and class. Capitalism exploits the intersectional differences that imperialism invented in new ways for the same purposes. In this text, we aimed to outline the many and various ways that imperialism continues to exploit intersectional others in the name of economic development.

In spite of the failures of neoliberal philosophy that came to light in the 2008 financial crisis and a variety of alternatives, globalization, driven by neoliberalism, continues to be the primary mode of state engagement with economic life. The continued rollback of state regulation for public goods as well as deepening inequality result in multiple crises affecting the continued capacity of people to live on the planet, such as climate disruption, lack of access to water and sanitation, infant and maternal mortality, and conflict and food insecurity. Benería, Berik and Floro (2015) point out three contradictions that are inherent to development as usual. The first is the continued **discursive** (or through the use of language) separation of poverty reduction from wider capitalist processes that produce it. This can result in poverty reduction programs, such as microcredit, which create new kinds of economic dependencies and do not address why people are poor in the first place. The second is the response to economic crises has not typically been to increase state spending on health, education and welfare. Rather, capitalism has been supported and protected by countries, including the United States, financially bailing out their banks in response to the 2008 financial crisis. Last, gender continues to be a key area of intervention in development aid and work, but women are increasingly invoked discursively as subjects in need of care, humanitarian assistance or market-based solutions. This approach does not allow that the production of poverty might be an indicator of a problem with the capitalist system, nor does it make room for the crafting of policy that might incorporate the needs and aspirations of those impacted most. For example, Afghan women were indeed oppressed under the Taliban regime (1996–2001), as were many Afghan men. The US-led invasion and occupation of Afghanistan included focusing on "saving" women from the Taliban and local patriarchy. However, efforts to assist/save these women fell short because they did not incorporate local voices or the diverse needs of women, or include Afghan women and key decision makers about their own futures.

Thus, development work needs to decouple itself from capitalism, both as a way of pursuing development (microcredit, structural adjustment, loans, new markets) and the financialization of development, which benefits the developers more than it benefits the recipients. Benería, Berik and Floro (2015, 238) suggest the way forward is "to pursue development strategies that transform the productive structures of economies so as to support livelihoods in a sustainable manner and systematically address issues of distribution." The World Social Forum (WSF) was organized in the early 2000s as a global scale social movement of groups and civil society organizations opposed to neoliberalism, neo-imperialism and globalization. The diverse array of participants in the events hosted by the WSF were united in their resistance to global capitalism and corporate control of the economy. They imagine new ways of organizing the economy as well as developing new modes and scales of citizenship (Conway 2004). The movement starts from the premise that people have the right to have rights, followed by the development of alternative notions of rights that include "previously unimagined rights that emerge through specific political struggles" (Conway 2004, 374). This new citizenship is not linked to prior political authorities, but emerges in shared responsibilities to ensure the rights of others in civil society. The concept of a social and solidarity economy (SSE) emerged through the WSF, which aims to create an alternative globalization focused on equitable and democratic social relations in the economy, including intersectional equity in natural resource distribution.

Another critical form of resistance has been through the development of **transnational feminist networks** (TFNs), which are broad coalitions of women who advocate for women's rights at a variety of scales (Ferree and Tripp 2006). They formed in large part to resist the implementation and consequences of structural adjustment policies in spaces of dispossession that disproportionately impacted intersectionally gendered subjects. They often act as nongovernmental organizations and organize to have a voice in global platforms. They include the aforementioned DAWN (Chapters 1 and 2) as well as other organizations working to turn local feminist concerns into global scale efforts. Moghadam (2005, 104) writes that TFNs are

> tackling both the particularistic and the hegemonic trends of globalization. They are advancing criticisms of inequalities and forms of oppression, neoliberal economic policies, unsustainable economic growth and consumption, and patriarchal controls over women. In a word, transnational feminist networks are the organizational expression of the transnational women's movement, or global feminism.

Another critical goal of TFNs is pursuing human rights through peace. Conflict and violence have particularly gendered dimensions and manifestations, including rape as a weapon of war, and TFNs aim to make women and men powerful actors in waging peace, not war.

We would like to use the remainder of this chapter to outline some ways development might produce political-economic situations that take seriously the well-being of women and people of color, and that do not come at the expense of others. As we have indicated in previous chapters, development, when it takes place as capitalist economic expansion, facilitates dispossession and immiseration, not well-being for all. Anti-capitalist politics abound in a diversity of places, and they offer alternative structures of production and distribution that challenge the ongoing neoliberal business as usual. Anti-capitalist or post-capitalist politics address how markets generate poverty and the subjects in need of development, and prop up development practice as capitalism in another guise. The chapter is structured loosely around the order of

Chapters 1–9, with a discussion of primary, secondary and tertiary sector work coming first, followed by interventions in security and health. While this chapter does not focus specifically on gender, each of the efforts we outline below includes equity among and for intersectional others as part of their social platforms, as well as an overt emphasis on equity in general. This chapter outlines alternative sociopolitical arrangements and economic-ecological alternatives, as well as resistances. We conclude with a discussion of political and economic change and new kinds of decision-making arrangements that we feel are central for the production and distribution of resources in a post-development and decolonized world.

Primary sector: appropriate technology

Having much in common with the World Social Forum, La Via Campesina (LVC) is a peasant-led international social movement composed of groups from 70 countries, formed to address issues of food insecurity and farming livelihoods. They have historically condemned states, corporations and supranational organizations for their failures to provide real and lasting food security for the world's poor. An early document lists six key elements of a model for an alternative food system, the first three of which propose to redress inequalities in access to productive resources such as land. LVC and activists for food sovereignty demand the physical spaces to grow and consume food, but also the political space for life (Trauger 2017). LVC aims to do this by defending "small-scale sustainable agriculture as a way to promote social justice and dignity. It strongly opposes corporate driven agriculture and transnational companies that are destroying people and nature" (LVC 2012). LVC is an "an autonomous, pluralist and multicultural movement, independent from any political, economic or other type of affiliation" (LVC 2012). LVC's local affiliates work on a variety of campaigns including anti-GMO activism, gender equity and agrarian reform (Desmarais 2012). More recent efforts towards increasing autonomy in the food system have resulted in political interventions, often led by women, to protect local agriculture, seed saving and community-based food security (Trauger 2017).

Many communities seek to take control of energy and food rather than leave it in the hands of the state or corporate monopolies. In the aftermath of the natural disasters of Hurricanes Irma and Maria in the Caribbean, many islands are considering rebuilding their electrical grids with solar and other alternative energy technologies. Obtaining energy for household use in the poorest communities is typically the work of women and girls, and usually involves the gathering of firewood. This can take many hours of a day, and in the context of climate change, resources are increasingly scarce. The use of coal and charcoal in cooking fires deteriorates indoor air quality, and women are those most frequently exposed to noxious chemicals while cooking or working. To address the problem of monopoly control of energy, detriments to women's time, energy and health, as well as resource scarcity, decentralized alternative energy technologies at the household and village scale have been introduced in a variety of contexts.

For example, as discussed in Chapter 8, the introduction of solar cookers and **biogas** stoves to rural villages in China saved women time and money, and improved health and well-being outcomes (Ding et al. 2014). Decentralization is key to achieving equitable outcomes from renewable energy investments, as well as an overt emphasis on distributing energy equitably. For example, a centralized solar park in India was meant to provide renewable energy in an equitable way to marginalized rural communities, but instead led to inequities in distribution that benefited those with power, and led to the accumulation of benefits at wider social scales (region and nation) than the household (Yenneti and Day 2016). Similarly, a community energy project in the UK that was designed to be controlled by a community who directly benefited required far more input from individuals in the community than was initially expected, as well as robust policy protections from the state to help the project grow (Seyfang et al. 2014). Also, a village-scale solar power center in Kenya that incorporates appropriate technology such as solar-powered lanterns and mobile phone charging services was far more effective in meeting local needs than a large- or even medium-scale grid system (Ulsrud et al. 2015).

Water resources are directly linked to both food and energy production, in terms of using water for production, but also because of the likelihood of pollution from both. Women's work in the zones of dispossession in rural places often involves the difficult, time-consuming and exhausting work of bringing water to homes. In the context of urban areas, potable water tends to be the most expensive household item after food expenses and must be purchased and brought to households over long distances. In the zones of accumulation, industrial pollution has reduced water potability and safety in many cities, while agricultural pollution has contaminated ground water in rural areas. As the climate changes, water scarcity increases in some places, while flooding occurs in others, and this contributes to water quality and safety issues. Access to safe and affordable water is a human right, but one that is becoming increasingly difficult to realize. Efforts to secure water resources from pollution intersect with an indigenous and peasant-led movement to keep fossil fuels underground to prevent the release of carbon dioxide and slow the rate of climate change (Osborne 2017).

Social movements around water include efforts to protect sensitive water resources from potentially polluting industrial infrastructure, such as the pipeline protests outlined in Chapter 4. There are social movements happening in nearly every place that petroleum infrastructure threatens sensitive ecological areas and water resources, when states enable and abet efforts to develop in ways that are damaging to people and resources, and do not provide good alternatives. Other efforts have joined renewable energy production to sanitation through the use of membrane-based filtration systems powered with renewable energy. These systems can remove biological and chemical contaminants at the same cost as untreated water systems. They also have the potential to "leapfrog" over larger-scale infrastructure projects that would bring water and electricity to remote places (Schäfer, Hughes and Richards 2014). These projects often require large investments of capital, bringing in outside influence through foreign direct investment, and, frequently, corruption (Perkins 2007). While the membrane filtration technology is just one example, it promises to provide a basic need for clean water at low cost, with appropriate technology that benefits the community more than it would benefit an outside developer.

Three themes emerge from these cases: 1) the idea that instrumental and extractive ideologies are universal belies the persistence of post-humanist orientations toward nature, such as discussed by Zoe Todd (2014) and Dawn Hoogeveen (2016); 2) the conventional, centralized and industrialized models of producing food, water and energy are frequently inexact and inefficient fits for the communities they serve; and 3) smaller, more flexible, more community-owned and -managed projects do more to meet needs equitably and maintain community autonomy than large-scale infrastructure projects favored by the overwhelming majority of development agencies and nonprofits. Development that puts people at the center, rather than profit at the center, has many multiplier effects, and few risks. Scale-appropriate and nonpolluting technology requires protection from corporate profit interests by the state, as well as incentives, rather than obstruction and inertia and bureaucratic approaches to providing basic services.

Secondary sector work: wage solidarity

Industrialization made manufacturing the bedrock of the capitalist economies of the newly emerging states during the 20th century. As globalization allowed capitalists to seek new labor markets and spaces of accumulation, the benefits of manufacturing increasingly accrued to global elites who now control the vast majority of the world's wealth. Marx (1867, 1977) wrote that capitalism alienates workers from each other, the production process and their product. Secondary sector work that provides a good income and safe working conditions and does not alienate workers is difficult to find in the global assembly line. Corporations that race to find employees at bottom wage levels require spaces of dispossession, rendering more and more people vulnerable to exploitation and displacement. There are some models of secondary sector work, however, that provide the means for a good livelihood, decision-making power and collective power within the enterprise.

The Mondragon Corporation is a worker-owned cooperative in the Basque country in Spain. It was founded in 1956 as a way to foster participation and solidarity in workplaces. The cooperative is organized around a few key principles of social justice, anti-capitalism and common goods. At the center of its value system is education and employee ownership. Derived from this are wage solidarity, participation in management, subordination of capital to labor, internal accountability among workers and nondiscrimination. **Wage solidarity** means that there is a small differential between the lowest-paid and the highest-paid worker. In the case of Mondragon, this ratio is 1:6. The subordination of capital to labor means that worker well-being is more important than profit. While it has not been without its problems, particularly during the 2008 financial crisis, the cooperatives were at the heart of some of Spain's only successful economic regions. During the hardest times of the economic crisis, while the rest of Spain endured 26 percent unemployment, the cooperatives maintained steady employment and production. The New Era Windows cooperative in the United States is another example of a worker-owned cooperative that is managed democratically, with similar principles to the Mondragon co-ops.

Caps on the salaries of the highest-paid individuals in an enterprise are a key element of Benería, Berik and Floro's (2015) ideas about people-centered development. A lack of wage solidarity leads to inequality, which leads to increasing exploitation and dispossession. For workers to have lives that are characterized by well-being and fairness, wage solidarity is key. This is especially significant for intersectional others, who are frequently singled out for lower pay. Iceland has made efforts to address gender inequities through tripartitism, a strategy that combines employer associations, trade unions and national states in wage and welfare negotiations, with mixed success (Fraile 2016). Another approach has been through transnational labor solidarity, but this also has met with mixed success due to differences in perspective on "free trade" and position relative to capital accumulation across the global assembly line (Bieler 2013). As argued by Mohanty (2003), for labor rights organizations to succeed, they must incorporate and ensure gender and race equality; and for feminist ideologies to succeed, they must insist upon racial, labor and class equity.

A theme emergent from these studies suggests that when equity is planned for (between classes, races or genders), it is more possible than when it is assumed to be handled by market forces. Faith in the ability of capitalism to provide public goods for all does not confront the way implicit and explicit bias operate to differentiate workers into hierarchies based on intersectional identities. Equity and solidarity must be planned and designed into systems and structures of employment and industry. An example of this would be planning for paid family leave at the state level with daycare facilities available in workplaces. The state has historically enabled wage differentials in an adherence to neoclassical and trickle-down economics, in which faith in accumulation in the private sector will lead to prosperity for the nation-state. Market fundamentalism has not worked, and all nation-states should craft policies that meet the needs of workers, regardless of their racial and gender identity or class position, and can facilitate class solidarity across borders.

The general themes we have identified above regarding decentralization, appropriate technologies, the subordination of capital, sustainability and well-being apply to these forms of development as well. Communication technologies in many spaces of dispossession appear without large-scale infrastructure projects, such as cellular and satellite technology "leapfrogging" land lines in many remote rural areas of Asia and Africa. Universal education is widely available in many countries, and we feel that expanding the scope, scale and democratic accountability of such opportunities will benefit all. Similarly, transportation infrastructure can be built to accommodate the most environmentally friendly and people-centric technologies, such as solar bike paths in the Netherlands or trinary road systems in Curitiba, Brazil, which accommodate bikes, buses, cars and people.

Tertiary sector: service work

Service work under conditions of late capitalism in zones of accumulation shares features with dystopian genre fiction. A recent article in the *New York Times Magazine* described the future of service work as similar to working for Amazon, the largest online retailer in the United States, which has been characterized as "unapologetically ruthless" – long hours for employees, invisibility to consumers and automation. In contrast to other attempts to make service work more transparent, equitable and fair, Amazon effectively erases workers and knowledge about working conditions, and consumers don't seem to mind (Herrman 2017). As Amazon searched for its second headquarters in 2017, it was courted by "right to work" states such as Georgia and Colorado, which prohibit collective bargaining.

Service work in spaces of dispossession remain largely in the informal sphere. Neither taxed, nor regulated by the state, they are spaces of precarity and vulnerability for women and intersectional others. Lack of formal recognition and protections make service work difficult and dangerous for many. Kate Swanson (2010) writes that the kinds of work available to indigenous women in Ecuador involve physical separation from their children, assault and exploitation, especially in informal domestic work, and discrimination in formal sector work. Paradoxically, informal work in public, such as selling gum on the street, allowed more autonomy and better life chances for their children. Service work is also done by entire countries. For example, spaces of dispossession are involved in providing environmental services, such as climate mitigation strategies. These often involve planting non-native species as carbon sinks or planting trees in grasslands and disrupting agroecological systems (Osborne 2017). In this way, those least responsible for climate change do the most work and bear the costs of mitigation. The knowledge sector is growing in spaces of dispossession, although they often experience "brain drains" of skilled workers to more lucrative labor markets, and the technology sector is contributing to the growth of a consuming middle class, which may not be ecologically sustainable.

An alternative way to provide services that persists across zones of accumulation and spaces of dispossession is the development and maintenance of **reciprocal social economies** (RSEs). These are systems of exchange, in which food or other goods and services are provided via ritualized giving or through reciprocal arrangements. A key component of RSEs is that goods are given and services are provided without explicit understanding that a similar good or

FOCUS: COOPERATIVE HOUSING LEADS PEOPLE-CENTERED DEVELOPMENT

In previous chapters we described the temporary and precarious housing situations that poor migrants often encounter when they arrive in a new country or a new city: slums, shanty towns, tent cities or high-rent but low-quality apartments or motels. A collective housing project in the city of Solapur in Maharashtra, India, indicates that worker cooperatives can pressure governments to help construct affordable, quality housing for intersectionally oppressed workers. The women who inspired this movement were low-wage workers, many from lower castes and religious minority groups without much formal education. After a long struggle waged with large trade unions across India, the workers won support from the government and construction began in 2001. The housing scheme includes 30,000 units approximately 555 square feet in size, open space, schools and hospitals. The workers purchased the housing units for the equivalent of $300, or one-third of the construction cost. People-centered development has followed: more workers joined, more schools were built, the hospital expanded and a bustling market (Revolution Square) was built. (For more information, see Tricontinental 2018.)

service will be given in exchange at some later date (Malinowski 1922). Reciprocation breaks down the way services have become commodified under conditions of capitalism. While it is widely thought that the commodity situation dominates most transactions, the transformation from premodern to modern economies is incomplete. According to White and Williams (2012, 1636), the world "should be more properly understood as a largely non-capitalist landscape composed of economic plurality, wherein relations are often embedded in non-commodified practices such as mutual aid, reciprocity, co-operation and inclusion."

Precapitalist modes of exchange abound, even as they are marked as backward or primitive by a state–market nexus concerned with the production and appropriation of surplus. Offer (1997, 457) calls these exchanges "relations of regard," "voluntary transfer," "a self-enforcing bond" and expectation of reciprocity, which is "motivated by a desire for regard, over and above any gains from trade." The essential feature of any economy is the way "transactions that surround things are invested with the properties of social relations" (Appadurai 1988, 15). The social relations around RSEs aim to provide the insurance, financial and welfare systems that in many cases the state or markets fail to provide for people. Anarchist political theory suggests that RSEs have the potential to resolve the contradictions of capitalism, particularly those that consolidate capital in the hands of a few, resulting in poverty for the many.

The role of the state

All of the above requires a different kind of state, if not different forms of democracy itself. Two themes emerge from an analysis of how capitalism works through a state that enables it: 1) productive activity is designed to accumulate surplus in the private sector; and 2) borders function as ways to control the population and make them subject to the will of state capitalism. A reworking of state power is part of the decolonization project, and thus decolonization must be linked to a broader set of issues regarding rights within the state system as well as the way in which capitalism shapes economic relationships between people within states as well as between states.

Population

Under the state system, the work of population management has historically been the state's responsibility. The role of the state was to manage the size, rate of growth, composition and increasingly the health of its citizens/workers. It is now increasingly scaled up to the supranational scale or down to the local scale. The supranational actors are frequently out of touch with the needs of women and families in terms of their decision making, and the local scale, if more in touch, is frequently under-resourced. Irrespective, the future of population growth policy must account for the ways in which access to healthcare, particularly maternal care, shapes population outcomes. This includes access to contraception, abortion, prenatal, maternity, postnatal and neonatal care. This is difficult and expensive to provide, but remote places with few resources have experimented with mobile maternity clinics to provide basic, lifesaving care. However, no decision about family planning policies can proceed without consideration of the way in which a woman's autonomy is frequently limited in terms of her sexuality, as well as how cultural values shape ideas about limiting (or not) family size. While the state remains responsible for its population, more decentralized, culturally sensitive and women-centric policies and approaches are required now more than ever.

Humanitarianism

The Rohingya refugee crisis in 2016–17 underscores the need for robust and well-funded international humanitarian relief efforts. The military and other armed groups of Myanmar ethnically cleansed a Muslim minority in the southern part of the state. The ensuing millions of refugees crossed the border into Bangladesh and as of this writing are desperately waiting for food and shelter. They included thousands of orphaned children, widows and other vulnerable women who are at acute risk of sexual exploitation and trafficking (Cochrane 2017). Aid agencies are stretched to their maximum capacity. While this is clearly a case where international aid is vital and necessary, we suggest new methods of conceptualizing humanitarianism and aid. 1) The production of difference along the lines of some cultural identifier is part of the production of inequality so central to imperialism and capitalism. The Rohingya are an exploited, stateless minority within the territory of Myanmar that was created through successive waves of imperial control in the region. 2) The sovereignty of states is a key part of the problem in Myanmar because it means that states may murder people within their own territorial bounds with impunity. This is not the only incident of its kind in recent history and is a problem that will only go away with the abolition of state sovereignty. 3) State monopoly control of territory and borders means that the Rohingya have no rights within the state system when they cross a border. They are at the mercy of their host state to grant them basic human rights. Without addressing these central causes of the production of stateless, vulnerable refugees, humanitarianism provides a temporary solution to a problem that will never go away as long as the liberal state system persists.

Decolonization

Decolonization movements aim to free indigenous people from the grip of imperialism and state capitalism, and their struggles have a wider resonance with social movements for justice, equity and sustainability, although they are not synonymous with them. Tuck and Yang (2012, 3) write that decolonization is not a metaphor, it is not an "approximation of other experiences of oppression. . . . Decolonization doesn't have a synonym." Decolonization means the abdication of settler futures and returns sovereignty over territory to natives and the indigenous. What happens after that is not currently known. Tuck and Yang say that it does not have to be known at the time of decolonization. They do say that until decolonization is achieved, **settler futurity** will continue to shape the way in which the world is formed and known. Even claiming land for a commons as advocated by food sovereigntists without decolonization perpetuates the erasure of "existing, prior and future Native land rights, decolonial leadership and forms of self-government" (Tuck and Yang 2012, 28). That is to say, none of what we suggest in this chapter can happen without decolonization.

Some final thoughts

After the revolution in Cuba, a basic income model was established that gave everyone a minimum income for work. It was meant as a form of wage solidarity in that the poorest paid workers did not receive disproportionately less than the highest paid (Pearson 1998). Cuba established a food ration system that ensured people did not go hungry and made state-held land available to anyone who wanted to grow food. These reforms were made in direct response to the way foreign multinationals impoverished and immiserated the most vulnerable people on the island – those working in the sugarcane fields – by enclosing common land and appropriating surplus for American capitalists. We argue that these kinds of reforms, along with democratic decision-making processes, could serve as a model for alleviating contemporary poverty, food insecurity, inequality and other development problems. Basic income grants are being trialed in a variety of countries, as well as at the scale of the city, such as the one currently being trialed in Stockton, California. No amount of accessibility to food will resolve the problem of hunger when buying food requires money in a situation when people have no money.

Raising capital and receiving credit for enterprises and innovation are essential for any economy. A decolonized and post-development economy will have to find ways to extend credit and raise capital in ways that are not exploitative and discriminatory. One method, often referred to as "love money," or capital raised by family and friends, has a new identity as crowdsourcing, and many digital platforms work to enable this process. Also, Islamic banking prevents nonmaterial forms of interest. Therefore, compounded interest (i.e., interest charged against interest already accrued) would be *haram* (forbidden) within Islamic banking.

Similarly, alternative currencies that are community-backed (not bitcoin-type currencies) are effective in building local capacities, collective welfare and a system of interdependent cooperation (Burke 2012). The barter systems and alternative currencies Burke investigated in a post-conflict Colombian city were designed and facilitated through alliances between the middle and working classes, peasants and others. Burke found that the opportunity to control more of the process of production and consumption gave the participants a significant sense of autonomy, but for the process to be successful, a social transformation of economic subjectivities was necessary. A simply anti-capitalist approach, often utilized by advocates of decommodification, does little on its own to change the social context within which poverty and hunger are (re)produced.

Another form of capital is land. LVC organizes itself around enabling access to collective land rights or usufruct rights. Use rights are antithetical to systems that rely on privatization and private property rights. Collective resource management is key to usufruct land rights, and also challenges state authority and control over land. Agarwal (2014) proposes that to improve food security outcomes as well as increase the expertise of farmers and the empowerment of women, land could be granted to women through land reform and resettlement, equitable inheritance laws and collective ownership through microcredit and communal schemes, such as cooperatives and land trusts. However, property regimes that favor some form of enclosure and centralized control, usually through government agencies, have not been beneficial to the commons, and run counter to decolonization efforts in which land should be returned to the indigenous inhabitants. Similarly, LCV emphasizes that indigenous land claims related to the commons must be recognized and "customary rights to territory must be recognized" (Nyéléni 2007, 56). Shared ecological capital, for which no one pays rent (much like land in the Cuban usufruct system), is the foundation of a noncommoditized and decolonized food system. However, to reclaim the commons in the context of the liberal sovereign state is paradoxical, unless new modes of governance can also be realized.

Central to the success of any of these efforts to decolonize and create alternatives to capitalism is **anti-oppression work**, which can provide the social and cultural transformation necessary for a transition to a more sustainable and equitable world. Raj Patel (2009, 671) writes in the context of gender equity and food sovereignty movements, "every culture must, without exception, undergo transformation." Anti-oppression work aims to transform the knowledge practices of societies founded on racial–sexual violence (Luchies 2015). Anti-racist pedagogies are essential to universal education that can facilitate a transition to decolonization as well. We take issue with "capabilities" approaches (Nussbaum 2001) that assume all people have the same life chances. When we live in a world that has been rendered fundamentally unequal through imperialism and capitalism, we all cannot realize our full capabilities. Without a social transformation that understands anti-colonial, anti-racist and anti-patriarchal regimes of truth as legitimate and beneficial to society, all social movements will remain mired in the violence of their social context. Ince (2012) suggests that prefigurative politics aims to build new political futures within the existing institutional frames of contemporary society. This must be done within the ecological and socio-economic constraints of the existing material world. Guiding its decision making, the World Economic Forum uses a schematic called the "doughnut" which illustrates how a future without inequality must happen within ecological limits and based on principles of equity and justice (Rockström et al. 2009). While all of what we have argued for in this chapter will take work, prefigurative politics assume that we can work where we are, that social transformation is never complete and that society is constantly in a process of becoming.

An intersectional gender approach to development indicates that all people are divided into groups along the lines of gender, race, class, sexuality, nationality, etc. This process is effective in identifying some as "worthy recipients of aid" in the form of humanitarianism (e.g., women refugees, but not able-bodied men from the same community) and/or who will be exploited on the basis of their lack of rights, citizenship or social personhood because of their identity so that capitalists can accumulate surplus in the private sector (e.g., women migrants working in the domestic sphere). We argue that both processes are tied to the same machinery of global capitalism. We agree with Žižek (2006) that "before you can give all this away you have to take it," meaning that humanitarianism is needed because governments, and empires before them, have taken capital and capacities through colonialism, capitalism and globalization. We argue that in order to remove the gendered intersectional oppression in the world today, we need to replace global capitalism with more just, inclusive and sustainable alternatives, and reform or dismantle the territorial-based forms of governance that support capitalism and craft policy to perpetuate it. We hope this book has provided the rationale for such a movement, as well as hope and inspiration for those who wish to wage peace, justice and sustainability in the world.

Recommended reading

No is not enough, Naomi Klein; *Disassembly required*, Geoff Mann

Recommended viewing

First daughter and the black snake; *Doughnut economics*

Questions for discussion

What does decolonization mean? Read Tuck and Wang (2012). What does decolonization mean for settler populations? How is inequality produced? How might it be ended? Go to this map: http://native-land.ca/. Find out who was here before you. How can decolonization be a shared process? How might decolonization affect you in your daily life? How can you make it happen? Watch the one-minute video *Doughnut economics* (www.youtube.com/watch?v=Mkg2XMTWV4g). Write a one-page response to the ideas presented.

References

Agarwal, B. (2014). Food sovereignty, food security and democratic choice: critical contradictions and difficult conciliations. *Journal of Peasant Studies*, 41(6), 1247–68.

Appadurai, A. (Ed.) (1988). *The social life of things: commodities in cultural perspective*. Cambridge: Cambridge University Press.

Benería, L., Berik, G., & Floro, M. (2015). *Gender, development and globalization: economics as if all people mattered*. New York: Routledge.

Bieler, A. (2013). The EU, Global Europe, and processes of uneven and combined development: the problem of transnational labour solidarity. *Review of International Studies*, 39(1): 161–83.

Burke, B.J. (2012). "Para que cambiemos"/"So we can (ex) change": economic activism and socio-cultural change in the barter systems of Medellín, Colombia. Unpublished PhD thesis, University of Arizona.

Cochrane, L. (2017). Human traffickers target Rohingya refugee camps. www.abc.net.au/news/2017-10-20/sexual-predators-human-traffickers-target-rohingya-refugee-camps/9068490. Accessed 10/26/2017.

Conway, J. (2004). Citizenship in a time of empire: the World Social Forum as a new public space. *Citizenship Studies*, 8(4), 367–81.

Desmarais, A.A. (2012). *La Vía Campesina*. New York, NY: John Wiley & Sons.

Ding, W., Wang, L., Chen, B., Xu, L., & Li, H. (2014). Impacts of renewable energy on gender in rural communities of north-west China. *Renewable Energy*, 69, 180–89.

Ferree, M.M., & Tripp, A.M. (Eds) (2006). *Global feminism: transnational women's activism, organizing, and human rights*. New York, NY: New York University Press.

Fraile, L. (Ed.) (2016). *Blunting neoliberalism: tripartism and economic reforms in the developing world*. New York: Springer.

Herrman, J. (2017). What will service work look like under Amazon? www.nytimes.com/2017/07/18/magazine/what-will-service-work-look-like-under-amazon.html.

Hoogeveen, D. (2016). Fish-hood: environmental assessment, critical indigenous studies, and posthumanism at Fish Lake (Teztan Biny), Tsilhqot'in territory. *Environment and Planning D: Society and Space*, 34(2), 355–70.

Ince, A. (2012). In the shell of the old: anarchist geographies of territorialisation. *Antipode*, 44(5), 1645–66.

La Via Campesina (LVC) (2012). http://viacampesina.org/en/. Accessed 6/5/2012.

Luchies, T. (2015). Towards an insurrectionary power/knowledge: movement-relevance, anti-oppression, prefiguration. *Social Movement Studies*, 14(5), 523–38.

Malinowski, B. (1922). *Argonauts of the Western Pacific*. London: George Routledge.

Marx, K. (1867, 1977). *Capital, vol. 1*, trans. Ben Fowkes. New York, NY: Vintage.

Moghadam, V. (2005). *Globalizing women: transnational feminist networks*. Baltimore, MD and London: Johns Hopkins University Press.

Mohanty, C. T. (2003). *Feminism without borders: decolonizing theory, practicing solidarity*. Durham, NC: Duke University Press.

Nussbaum, M.C. (2001). *Women and human development: the capabilities approach*. Cambridge: Cambridge University Press.

Nyéléni (2007). Proceedings of the Forum for Food Sovereignty, Selengue, Mali, February 23–27.

Offer, A. (1997). Between the gift and the market: the economy of regard. *Economic History Review*, 50(3), 450–76.

Osborne, T. (2017). Public political ecology: a community of praxis for earth stewardship. *Journal of Political Ecology*, 24(1), 843–860.

Patel, R. (2009). What does food sovereignty look like? *Journal of Peasant Studies*, 36(3), 663–706.

Pearson, R. (1998). The political economy of social reproduction: the case of Cuba in the 1990s. *New Political Economy*, 3(2), 241–59.

Perkins, J. (2007). *The secret history of the American empire: economic hit men, jackals, and the truth about global corruption*. New York: Penguin.

Rockström, J., Steffen, W., Noone, K., Persson, Å., Chapin III, F. S., Lambin, E. F., ... & Nykvist, B. (2009). A safe operating space for humanity. *Nature*, 461(7263), 472.

Schäfer, A.I., Hughes, G., & Richards, B.S. (2014). Renewable energy powered membrane technology: a leapfrog approach to rural water treatment in developing countries? *Renewable and Sustainable Energy Reviews*, 40, 542–56.

Seyfang, G., Hielscher, S., Hargreaves, T., Martiskainen, M., & Smith, A. (2014). A grassroots sustainable energy niche? Reflections on community energy in the UK. *Environmental Innovation and Societal Transitions*, 13, 21–44.

Swanson, K. (2010). *Begging as a path to progress: indigenous women and children and the struggle for Ecuador's urban spaces*. Athens, GA: University of Georgia Press.

Todd, Z. (2014). Fish pluralities: human–animal relations and sites of engagement in Paulatuuq, Arctic Canada. *Études/Inuit/Studies*, 38(1–2), 217–38.

Trauger, A. (2017). *We want land to live: making political space for food sovereignty*. Athens, GA: University of Georgia Press.

Tricontinental: Institute for Social Sciences (2018). The story of Solapur, India: where housing cooperatives are building a workers' city. www.thetricontinental.org/wp-content/uploads/2018/07/180704_Dossier-6_EN_Final.pdf. Accessed 10/8/2018.

Tuck, E., & Yang, K.W. (2012). Decolonization is not a metaphor. *Decolonization: Indigeneity, Education & Society*, 1(1), 1–40.

Ulsrud, K., Winther, T., Palit, D., & Rohracher, H. (2015). Village-level solar power in Africa: accelerating access to electricity services through a socio-technical design in Kenya. *Energy Research & Social Science*, 5, 34–44.

White, R.J., & Williams, C.C. (2012). The pervasive nature of heterodox economic spaces at a time of neoliberal crisis: towards a "postneoliberal" anarchist future. *Antipode*, 44(5), 1625–44.

Yenneti, K., & Day, R. (2016). Distributional justice in solar energy implementation in India: the case of Charanka solar park. *Journal of Rural Studies*, 46, 35–46.

Žižek, S. (2006). Nobody has to be vile. *London Review of Books*. www.lrb.co.uk/v28/n07/slavoj-zizek/nobody-has-to-be-vile. Accessed 10/7/2018.

Index

Locators in *italics* refer to figures and tables.

9/11 terrorist attacks 134

abortion 106, 107–108
accumulation: capitalism 14–15; colonialism 27; by dispossession 6, 7, 51; flexible 69, 74; and gender 7–8; *see also* zones of accumulation
added value, fair trade as 44
Afghanistan: disaster assistance 134–136, 142; sexual relationships 131
Africa, history under imperialism 19–25, *20*
Agarwal, B. 59, 149
agrarian crisis 68
agricultural modernization 68
agriculture: alternative development 143–144; biotechnologies 120–121; capitalist-export orientation approach 8–9; development as dispossession 51–52, 53, 58–59; enclosure 26; fair trade 44–46; gendered access to land 58–59; Green Revolution 8, 14, 56, 102–103; land grabs 61–62; Mexico case study 77; pesticides 14, 61, 102, 121; population concerns 102–103; as primary economic activity 55–56; single resource economies 54–55
aid: and development 10–11, 35–39; development workers 37–39; disaster assistance 127–128; disaster capitalism 128–129; 'needing help' 15, 30, 35; *see also* humanitarianism
alcohol abuse 97, 106
All India Women's Conference (AIWC) 106–107
alternative currencies 149
alternative development 141–143, 148–150; DAWN 13–14, 142; manufacturing sector 144–145; primary sector 143–144; role of the state 147–148; service sector 146–147
Amazon 146
Ambedkar, B. E. 113
Anderson, Warren 14
antenatal care 98–99
anti-capitalism 142, 149
anti-oppression work 149
artisanal and small-scale mining (ASM) 62
Australia, refugees case study 55
authenticity work 117
automation 112

Bacon, Francis, Sir 24
banana industry 44–46
Bangladesh, Rohingya refugee crisis 147–148
barter systems 149
basic income model 148
bathrooms and sanitation 96
beauty, perceptions of 120
'beltway bandits' 36–37
Benería, L. 141, 142, 145
Berik, G. 141
bin Laden, Osama 134
binary thinking 20, 85, 86
biogas 122–124, 143
biotechnologies 117–122
birth control 103–104, 107
birth rates 100, 106, 107
black feminism 4
blood quantum 23
body shopping 117
Bono 41
Boserup, Esther 13, 102–103, 113
Bretton Woods agreement 6, 27–28
British Empire 27, 32; *see also* colonialism
Buckley, M. 76
Buddhists 104
business for international development *see* development businesses
business process outsourcing (BPO) 116–117
Butler, Judith 12, 38

Cáceres, Berta 61
Cáceres, Dolores 141
call centers 116–117
Cambodia, technology case study 119–120
capitalism: alternative development 141–142; anti-capitalism 142, 149; and development 5–6; development as dispossession 53–55, 58, 63; and gender 30; inequality 3–4, 58; infrastructure development 66; intersectional approach 14; intersectionality and feminism 30–32; land dispossession 141; mixed economies 72; nation states 25–30; quality of life 14–15; wage solidarity 145
capitalist development 6–7
capitalist-export orientation approach 8–9; development pathways 68; humanitarianism 10; primary economic activities 54
careers in development *see* development workers

caste 12, 113
categories of difference 32
celebrity advertising 39–42, 46
central planning 9–10
charity organizations 36, 39–42
chemical leak, Union Carbide 14
chemicals in the agricultural industry 121
child sexual abuse 131
childbirth 98–99
childcare services 83, 84
China: communism 27; economic development 7; one-child policy 104
Chipko movement 60
chocolate industry 3
circular migration 70
cisgender 12, 73
class: biotechnologies 117–119; intersectionality 12, 31; middle-class women in work 83; migrants 86, 89–90
climate change 63, 144
cocoa plantations 3
Cold War 6, 7, 27–28, 29
collective housing projects 146
collective land rights 149
colonial debt 28
colonialism: accumulation of capital 27; decolonization 5, 147, 148; and development 6; development as dispossession 52–53; and empire 21–25; exploitation 32; financial 7; gendered access to land 58–59; and humanitarianism 10; infrastructure development 66–67; primary economic activities 54, 55; technology 113–114
commercial farming 56; *see also* agriculture
commodification of celebrities 41
communism 27, 105, 106
compulsory heterosexuality 12
conflict: Afghanistan 134–136; disaster assistance 127–128; disaster capitalism 128–131; displacement 133–134; future challenges 137–138; humanitarian militaries and liberal peace building 131–134; and migration 78–79; Syria 136–137
consumption: development pathways 69; fair trade 44–46; morality 40
contraception 103–104, 107
contract work, ICT professionals 117
cooperative housing 146
Council of Popular and Indigenous Organizations of Honduras (COPINH) 61
Court, G. 85
creative economy 90
credit: and humanitarianism 11; micro-credit 11, 42–44, 43–44; post-World War era politics 28; single resource economies 54–55
criminal work, informal economy 75
crowdsourcing 148
Cuba: basic income model 148; as planned economy 71; post-World War era politics 28; primary economic activities 54
Czech Republic, population 105–106

Dakota Access Pipeline (DAPL) 51, *52*
dalit 113
Darwinism 21–22, 24–25
de Condorcet, Marquis 21
death rates 100
debt: and humanitarianism 11; micro-credit 11, 42–44, 43–44; post-World War era politics 28; single resource economies 54–55
decentralization 143
decolonization 5, 147, 148, 149
Democratic Republic of Congo (DRC): child sexual abuse 131; development as dispossession 62; primary economic activities 54
Denmark, maternity/paternity leave 84
dependency 9–10, 54
deregulation 29
development 5–8; big D and little d 5; and colonialism 6; definition 5; discourses of 72–76; exacerbating inequality 15; and gender throughout history 13–15; and humanitarianism 10–11, 35–39; inequality 5; measurement 9; nation states 25–30; pathways to 68–72; technology and gender 112–114; theories and practices 8–10
Development Alternatives with Women for a New Era (DAWN) 13–14, 142
development as dispossession 51–53; agriculture, forestry and mining 58–61; capitalism 53–55, 58, 63; Democratic Republic of Congo case study 62; enclosure and privatization 53; future challenges 62–64; intersectionality 53–55, 63; primary economic activities 53–61; United States case study 61–62
development businesses: celebrity advertising 41; humanitarianism 36; micro-credit 43
development status, terminology 8
development workers: and aid 37–39; qualifications 38–39; salaries 38–39; sexual relationships 131
disaster assistance 127–128; Afghanistan 134–136, 142; future challenges 137–138; humanitarian militaries and liberal peace building 131–134; Syria 136–137
disaster capitalism 128–131
discrimination, intersectional 121–122
discrimination cases 81
discursivity, poverty reduction 141
displacement 68, 133–134
disposability of workers 73, 74, 77
dispossession *see* accumulation by dispossession; development as dispossession; spaces of dispossession
division of labor *see* gendered divisions of labor; new international division of labor
'docility', female workers 74, 77, 114–115
Dominican Republic: fair trade 44–46; forestry 60
double movement 29
drinking water 95–96, 144
drug abuse 97, 106
Dubai, UAE, labor case study 76–77

ecofeminism 4, 14
economic crises *see* financial crisis (2008)
economic development institutions 6
economic development programs 36; *see also* humanitarianism
economic development strategies 6–7
economic liberalism 25
education: planned economies 70; sanitation 96
Ehrkamp, Patricia 88
emancipatory technologies 122–124
embodied experience 4
embodied work 85
emotional labor 85, 86
empire, historical context 19–25
employment *see* development workers; labor

enclosure 26, 51, 53, 148
energy sources 122–124, 143–144
Engels, Friedrich 27
engendering 4
engineering 29, 86, 114
English language proficiency 38
Enlightenment: historical context 21; intersectionality 24; orientalism 21, 22
entrepreneurialism, micro-credit 43
environmental issues: alternative development 141; climate change 63, 144; water access 95–97, 144
environments, health in workplaces 95–97
equality: Cuba 71; fair trade 44–46; feminism 30–31; pay gap 72, 73, 81, 82, *83*; technology 114
ethical economics 43
ethnicity, intersectional approach 12
eugenics 21–22, 24–25, 103–104
Europe: capitalist economic philosophy 29–30; categories of difference 32; colonialism and the Cold War 6; economic development 7; empire in historical context 19–25; post-World War era politics 27–30
exploitation: colonialism 32; waged labor 73
exports *see* capitalist-export orientation approach
extraction *see* resource extraction

fair trade, international development 44–46
Fanon, Franz 24
farming *see* agriculture
Female Engagement Teams (FETs) 132
femicide 78
femininity 11–12, 62
feminism: disaster capitalism 128–129; intersectionality and capitalism 30–32; materialist feminist approach 4; relational thinking 20; rights 38; technology and gender 113; transnational feminist networks 142
feminisms (plural) 4
feminization of labor 8, 73–74, 90
fertility rates 100, 103, 105
feudalism 23
finance *see* credit
financial colonialism 7
financial crisis (2008): alternative development 141; capitalism 78; Dubai 76; Iceland 72; Indonesia 88; migration 76
financialization 51
First World 6
fisheries 56–57
flexibility in work 90
flexible accumulation 69, 74
Floro, M. 141
food and population 102–103
food preparation services 83
food safety 122
food sovereignty 63, 143
forced labor 70–71
forced migration 68
foreign direct investment (FDI) 69, 114
forestry: gendered livelihoods and resistance 60–61; as primary economic activity 57
for-profit government 78–79
free markets: and humanitarianism 10; market economies 68–69
free trade 145
fundraising, international organizations 40

Gandhi, Indira 107
Gates, Bill 58
gender: and accumulation 7–8; agriculture 56; biotechnologies 117–119; definition 4, 11–13; and development throughout history 13–15; feminization of labor 73–74; forestry 60–61; Germany migration case study 87–88; health and work environments 95–97; healthcare 98–100, 102; Indonesia case study 88–89; intersectional approach 5, 12; and knowledge economy 85–86; labor and exploitation 73; land access 58–59; micro-credit 42–43; migration and work 86–87; military campaigns 132–133; 'needing help' 30; performativity 12, 38; refugee resettlement 134; as relational and intersectional 11–12, 15; sex-selective abortions 107–108; technology and development 112–114; and the United Nations 29–30; vulnerabilities during conflict 129–130; wage equality 72, 73, 81, 82, *83*
gender advisors/experts/specialists 38
Gender and Development (GAD) 13
gender imbalance, Dubai case study 76, *77*
gender roles 11–12; access to land 58–59; household work 82; service sector 85
gender-based violence 129–131; *see also* violence against women
gendered divisions of labor 13, 117
gendered impacts of development 4
gender-equal pay 44–46, 73, 81, 82, *83*
General Agreement of Trade in Service (GATS) 118
genetically modified seeds 121
geographical context 15; healthcare 98; primary sector 54
Germany: maternity leave 84; migration case study 87–88
Ghana, water access 96
Giles, W. 133
glass ceiling 86
global assembly line 9, 73, 74, 81
global power 6, 7
Global Witness 141
globalization: migration and health 95; outsourcing 84, 89, 116–117; technology 114–115; theories and practices 9–10; Washington Consensus 29
Google 81
governments *see* state
Grameen Bank 42
Green Revolution: capitalist-export orientation approach 8; ecofeminism 14; implementation 56; population concerns 102–103
gross domestic product (GDP) 9
gross national product (GNP) 9
Grosz, E. 85

Haiti: 2010 earthquake 128; colonial debt 28; fair trade workers 46
Hartsock, N. 7–8
Harvey, David 26, 51
health insurance 99
healthcare: Czech Republic 105–106; future challenges and opportunities 109–110; gender, work and environments 95–97; India 106–109; labor environments 97; organ donors 118–119; planned economies 70; and population 95, 100–104; reproductive health 98–100, 102, 103–104, 105–106, 147

Heifer International 39
heteronormativity 12
heterosexual matrix 12
Hindus 104
HIV/AIDS 99
Honduras, forestry 61
Hoogeveen, Dawn 144
house searches 135–136
household work 82, 84
housing: collective housing projects 146; informal economy 74–76; planned economies 70–71
Huang, S. 86, 87
Human Terrain System (HTS) 132–133
humanitarian assistance programs 36
humanitarian militaries 131–134
humanitarianism: celebrity advertising 39–42; and development 10–11, 35–39; disaster assistance 127–128; disaster capitalism 128–129; fair trade 44–46; future challenges 46; intersectionality 149–150; micro-credit 42–44; Rohingya refugee crisis 147–148
Hurricane Katrina 127–128
hydroelectric power 114
hygiene 96–97
Hyndman, J. 128–129, 133

Iceland: financial crisis 72; pay gap 82; tripartitism 145
identity: intersectional gender 15; relational thinking 20
imperialism 4; categories of difference 32; definition 20–22; historical context 19–25; infrastructure development 66
import substitution 69, 77
India: caste 113; chemical leak 14; collective housing project 146; forestry 60; information and communication technologies 116–117; infrastructure and labor case study 66–67; land access 59; maternity leave 84; micro-credit 43; migration from 76–77; miscegenation laws 23; organ donors 119; population case study 106–109; population control 104, 105; reproductive technology 117–118; solar energy 143; technology access 116
indigenous populations: Dakota Access Pipeline (DAPL) 51, *52*; decolonization 148; development as dispossession 51–52, 63–64; gendered access to land 59; seizure of land *52*
Indonesia migration case study 88–89
Industrial Revolution 25, 27
industrialization 144–145
inequality: capitalism 3–4, 58; capitalist development model 58; and celebrity advertising 41; development 5; future challenges 46; Green Revolution 56; healthcare 95, 98–99; Sustainable Development Goals 3; uneven development 81–82, 87; wage gap 72, 73, 81, 82, *83*
infant mortality 100, 107
informal economy: housing 74–76; migrants 66–67; vulnerability and precarity 66, 84–85; women's work 88–89
information and communication technologies (ICT) 115–117
infrastructure development 66–67, 113
institutions *see* economic development institutions
internally displaced persons (IDPs) 68, 133–134
international development 35; celebrity advertising 39–42; fair trade 44–46; future challenges 46; humanitarianism and economic development 35–39; micro-credit 42–44
International Labour Organization 75
International Monetary Fund (IMF): formation 6, 28; population control 107; Washington Consensus 7, 29
international organizations: disaster capitalism 129; formation 6, 28; fundraising 40; healthcare 100; humanitarianism 36–37; population control 107; Washington Consensus 7, 29
intersectional discrimination 121–122
intersectional gender 15
intersectionality 4; capitalism 14; development as dispossession 53–55, 63; disaster capitalism 128–129; and feminism 31–32; feminism and capitalism 30–32; gender 5, 12; global economy 78; healthcare 99; humanitarianism 149–150; migration 89; 'needing help' 30; orientalism 21; poverty reduction programs 4–5; relational thinking 20; service sector 84–85
intersex 12
intimate geopolitics 104
Islam 88, 89, 104, 137
Islamic banking 148–149
isolationist trade strategies 9–10
Israel-Palestine conflict 104

Japan, population pyramid 100, *101*
Jolie, Angelina 41

Kabeer, Naila 4
Kenya: solar energy 143–144; technology case study 116, 120–122
Keynesianism 29, 78
Klein, Naomi 128
Kleiner Perkins 81
knowledge economy: discrimination 81; and gender 85–86; trends 82, 82–85

La Via Campesina (LVC) 143–144
labor: agriculture 58–59; development pathways 68–72; disaster assistance 135–136; discourses of development 72–76; Dubai, UAE 76–77; fair trade 44–46; female worker perceptions 74, 77, 114–115; feminization of 8, 73–74, 90; forestry 60–61; future challenges 78–79; gendered divisions of labor 13, 117; Germany migration case study 87–88; health and environments 95–97; Indonesia case study 88–89; informal economy 75–76; informal economy and vulnerability 66; knowledge economy 81; market economies 68–69; Mexico 77–78; migration 66, 68; migration and gender 86–87; mining 59–60; mixed economies 71–72; new international division of labor 81–82, 87; planned economies 70–71; service sector 9, 81; social reproduction 58, 105; technology 114–115; turnover rates 78; waged 73; Women in Development 13
labor solidarity 145
ladder model of development 69
land dispossession 141
land grabs 26, 51–53, 61
land mines, Cambodia 120

land ownership: alternative development 149; enclosure 26, 51, 53, 148; and gender 26; gendered access to land 58–59
legal context: deregulation 29; discrimination cases 81; food safety 122; micro-credit 43; miscegenation laws 23; one-child policy 104; reproductive technology 117; vulnerabilities during conflict 129–130
lesbian, gay, bisexual, trans and intersex (LGBTI) 137
less developed countries, terminology 8
liberal logic 30
liberal peace 133
liberal-capitalist feminism 4
life expectancy 100
Loewenstein, Anthony 78
Louis Vuitton Core Values campaign 41
love money 148

Machiavelli, Niccolò 21
male sterilization 107
male supremacy 4
Malkki, L. H. 37
Malthus, Thomas 102
manufacturing sector: alternative development 144–145; economic liberalism 25; female worker perceptions 74, 77, 115; technology 113, 114; three-sector theory 53; work environments 97
maquiladoras 114–115, *116*
market economies: labor 68–69; Singapore 70
market fundamentalism 69
Marx, Karl 27, 78, 144
Marxist-feminist approach 13, 78
masculinity: agriculture and ideologies 62; historical context 22–23; meaning of gender 11–12; and mining 59–60; sexually transmitted diseases 99
materialist feminism 4
maternal mortality 99, 100, 107
maternity leave 83, *83*, 84
Mayer, T. 104
McDowell, L. 85
McKittrick, K. 22
mercantilism 25
metropoles 23, 25
Mexico: labor case study 77–78; maquiladoras 114–115, *116*
micro-credit 11, 42–44
middle-class: agents of care 40; knowledge economy 86, 146; migrants 86, 89–90; women in work 83
migration: agricultural modernization 68; and conflict 78–79; displacement 133–134; economic crises 76; future challenges 89–90; Germany case study 87–88; healthcare 95; Indonesia case study 88–89; informal economy and vulnerability 66–67; meaning of 68; Singapore 70; work environments 97
Millennium Development Goals (MDGs) 3, 30
minimum wages 82–83
mining: Democratic Republic of Congo (DRC) 62; and masculinity 59–60; as primary economic activity 58
Mirchandani, K. 117
miscegenation laws 23
mixed economies 71–72
modernity: agriculture 56; Enlightenment 21; and technology 114; unequal social relations 53
Moghadam, V. 38, 142
Mondragon Corporation 145
Monsanto 121
morality: celebrity advertising 40–41; consumer purchases 40
more developed countries, terminology 8
multinational corporations (MNCs): definition 4; inequality 3; mining 58; *see also* transnational corporations
Muslims 88, 89, 104, 137
Myanmar, Rohingya refugee crisis 147–148

nation state, emergence of contemporary form 25–30; *see also* state
nationality, intersectional approach 12
natural disasters 127–129
nature: post-humanism 144; and technology 112–113
Nauru, refugees case study 55
Naybor, D. 59
Nazis 104, 105
neediness: and celebrity advertising 41; ‘needing help’ 15, 30, 35
neocolonialism, and humanitarianism 10, 11
neoliberalism 30, 127, 141
neo-Malthusians 102
New Era Windows cooperative 145
new international division of labor (NIDL) 81–82, 87; *see also* global assembly line
New Zealand, refugees case study 55
nonbinary gender 12
nongovernmental organizations (NGOs): biotechnologies 120–121; humanitarianism 36–37; Washington Consensus 7
Non-Profit Industrial complex 10
North American Free Trade Agreement (NAFTA) 77
North Korea 72

offshoring 84
one-child policy, China 104
oppression in feminist analyses 4
organ donors 118–119
organizations *see* development businesses; international organizations
orientalism 21, 22
Ottoman Empire 23–24, 27
outsourcing 84, 89, 116–117
Oxfam, future challenges 46

Pande, Are 118
Pao, Ellen 81
participatory planned economies 70
Patel, R. 116
patriarchy 4, 31
pay gap 72, 73, 81, 82, *83*; *see also* salaries
payday loans 43
peer learning 62
people-centered development 63, 79, 145, 146
performativity, and gender 12, 38
pesticides 14, 61, 102, 121
phenotypes 21–22
philanthropists 39–42
pipeline case study, Dakota Access Pipeline 51, *52*
planned economies 9–10, 70–71
Polanyi, Karl 29, 72, 78
population: Czech Republic 105–106; Dubai case study 76, *77*; future challenges and opportunities 109–110; growth and management 100–104; and healthcare 95, 100–104; India 106–109; reproductive health

98–100; state's role 147; women in waged work 83
population pyramids *77*, 100, *101*
postcolonial feminism 31–32
post-humanism 144
post-structuralist approach 5
poverty capital 11
poverty reduction programs 4–5, 141
precarity 84–85
pregnancy 98–99, 122
prenatal care 98–99
primary sector: agriculture, forestry and mining 58–61; alternative development 143–144; development as dispossession 53–58; geographical context 54; three-sector theory 53
privacy and sanitation 96
private sphere, service work 82
privatization: development as dispossession 53; sanitation 97
(PRODUCT) RED label 39–40
productive work, historical context 23
professionalization in development 38–39
professions *see* development workers
Project Everyone (UN) 3
property rights 66, 149
protectionist trade strategies 9–10
purdah 43

qualifications, development workers 38–39
quality of life 14–15
quaternary sector 53

race: biotechnologies 117–119; eugenics 104; healthcare 99; intersectionality 12, 31–32; knowledge economy 81; migration and work 86–87, 88; miscegenation laws 23; natural disasters 127–128; orientalism 21, 22
race to the bottom 69, 84
Radcliffe, S. A. 59
radical feminism 4, 31
rate of natural increase (RNI) 100
reciprocal social economies (RSE) 146
RED label 39–40
refugees: displacement 133–134; primary sector case study 55; Rohingya refugee crisis 147–148; Syria 136; and terrorism 89
regulations *see* legal context
relational thinking 20
reproductive health 98–100, 102, 103–104, 105–106, 147
reproductive technology 117–118
republics 25
reserve army of labor 73, 78
resistance, forestry 60–61
resource extraction: development as dispossession 51–52, 63–64; mining 58, 59–60, 62; primary sector 54
rights: feminism 38; food sovereignty 63; transnational feminist networks 142
Robbins, P. 103
robotic technologies 112
Rohingya refugee crisis 147–148
Rostow, W. W. 69, 74
Roy, Ananya 43, 62–63, 76
Rubin, Gayle 85

safe spaces 117
Said, Edward 21
salaries: development workers 38–39; gender-equal pay 44–46, 73, 81, 82, *83*; planned economies 70; wage gaps 73, 81, 82, *83*; wage solidarity 144–145
Sanger, Margaret 103–104, 107
sanitation 96–97
science, and the Enlightenment 24–25
Second World 6
secondary sector *see* manufacturing sector
self-sufficiency 9–10
service sector: alternative development 146–147; discrimination 81; emergence of 9; future challenges 89–90; and gender 85–86; meaning of 82; three-sector theory 53; trends 82–85; wages 9, 82
settler futurity 148
sex: and gender 12; imperialist age 23
sex work 85–86, 131
sexism in healthcare 98–99
sex-selective abortions 107–108
sexuality: and gender 12; intersectionality 31–32
sexually transmitted diseases 99
'shock city' 76
Sierra Leone, population pyramid 100, *101*
Silvey, Rachel 88–89
Singapore: as market economy 70; migration and work 86–87
single resource economies 54–55
skills, technology transfer 112
SKS 43
slavery: contemporary context 22; Enlightenment 21–22; and gender 8; historical context 19–25
slums 66, 75–76
Smith, S. H. 103
smoking 106
social and solidarity economy (SSE) 142
social biology 21–22, 24–25, 104
social class *see* class
social context: gender as 85; unequal social relations 53
social contract 26
social personhood 59–60
social reproduction 58, 85, 105
socialism, Czech Republic 105, 106
socialist feminist approach 13
socially reproductive labor 58
soft power 127
solar energy 143–144
sovereignty: state 26; workers 79
Soviet Union: Cold War 6, 7; communism 27; international aid 127; Washington Consensus 29
spaces of dispossession 8, 84, 89
Spain, Mondragon Corporation 145
special economic zones (SEZs) 69
Special Investigator General for Afghanistan Reconstruction (SIGAR) 135
Sri Lanka, 2004 tsunami 128–129
star/poverty space 41
state: alternative development 147–148; emergence of contemporary nation state 25–30; healthcare 100, 109–110; mixed economies 72; population control 104; primary economic activities 55; social reproduction in the Czech Republic 105, 106; sovereignty 26
stem cell research 119
STEM fields (science, technology, engineering and math) 29, 86
sterilization 107
structural adjustment programs (SAPs) 7
structural violence 41, 103
sub-Saharan Africa, HIV/AIDS 99
subsidies 70
subsistence farming 56; *see also* agriculture
suffrage movements 30
sugar industry 25

superpowers, global influence 6
supply chains 9
supranational organizations 4; *see also* international organizations
surrogacy, reproductive technology 117–118
Sustainable Development Goals 3
Swanson, Kate 146
Syria, disaster assistance 136–137

the Taliban 134, 135, 142
tea industry 25
technology: agriculture 56; alternative development 143–144; biotechnologies 117–122; Cambodia 119–120; emancipatory 122–124; gender and development 112–114; globalization 114–115; Green Revolution 56; information and communication 115–117; Kenya 120–122; military campaigns 132–133; population control 103–104
technology transfer 112
terrorism: 9/11 134; and migrants 89; War on Terror 7, 27, 132, 137
tertiary sector *see* service sector
Third Way (mixed economies) 71–72
Third World 6, 13
three-sector theory 53
tobacco use 106
Todd, Zoe 144
total fertility rates 100, 103, 105
tradition vs. modern dichotomy 114
trafficking 130–131
transgender 12
transnational corporations (TNCs) 114; *see also* multinational corporations
transnational feminist networks (TFNs) 142
Treaty of Westphalia 25–26
Turkey, migrants from 88
turnover rates 78

Uganda: healthcare 99; land access 59
underdevelopment 87
unemployment 73, 88
uneven development 81–82, 87
Union Carbide pesticide leak 14
unions 75
United Arab Emirates 76–77
United Kingdom, solar energy 143
United Nations Security Council (UNSC) 129–130
United Nations (UN): 2017 General Assembly 3; disaster capitalism 128; formation 28–29; and gender 29–30; Sustainable Development Goals 3
United States: Amazon 146; 'beltway bandits' 36–37; Cold War 6; Dakota Access Pipeline (DAPL) 51, *52*; development as dispossession 61–62; Hurricane Katrina 127–128; hydroelectric power 114; land grabs 61; maternal mortality 99; micro-credit 43; military campaigns 132–133; New Era Windows cooperative 145; refugee resettlement 134
United States Agency for International Development (USAID) 11
universal healthcare 109–110
USAID 121, 127
use rights 149
use values 85
usufruct rights 149

van der Ploeg, J. 73
Venezuela, primary economic activities 54
violence against women 78, 97, 129–131
voluntary migration 68
vulnerability: disaster assistance 129; informal economy 66, 84–85; Syrian refugees 136–137

wage gap 72, 73, 81, 82, *83*; *see also* salaries
wage solidarity 144–145
waged labor 73; *see also* labor
War on Terror 7, 27, 132, 137
Washington Consensus 6–7, 29
water access 95–97, 144
Western world: capitalist economic philosophy 29–30; categories of difference 32; colonialism and the Cold War 6; economic development 7; empire in historical context 19–25; post-World War era politics 27–30
Westphalia Treaty 25–26
White, R. J. 147
Williams, C. C. 147
'womb wars' 104
women: access to land 58–59; as caregivers 40, 41; and development throughout history 13–15; disaster assistance 128, 129–132, 134–136; fair trade 44–46; female worker perceptions 74, 77, 114–115; humanitarianism and aid 11; labor and exploitation 73; in mining 60; reproductive health 98–100; *see also* gender
Women, Environment and Development (WED) 13–14
Women and Development (WAD) 13
Women in Development (WID) 13
women's movements 30; *see also* feminism
worker sovereignty 79
workers *see* development workers; labor
World Bank: formation 6, 28–29; mining 58; population control 107; Washington Consensus 7, 29
World Economic Forum 149
World Health Assembly 118–119
World Health Organization (WHO) 100–102
World Social Forum (WSF) 142
Wright, M. W. 73–74, 78

Yakini, Malik 63
Yeoh, B. S. 86, 87
Yunus, Muhammad 42

Zelaya, Manuel 10
zones of accumulation 8; humanitarianism 35; migration 89; service sector 146